The
All-American
Map

Published for the Hermon Dunlap Smith Center for the History of Cartography at the Newberry Library

The
All-American
Map

Wax Engraving
and Its Influence
on Cartography

David
Woodward

The University
of Chicago Press
Chicago and London

DAVID WOODWARD is Director of the
Hermon Dunlap Smith Center for the
History of Cartography at the New-
berry Library, Chicago.

THE UNIVERSITY OF CHICAGO PRESS, CHICAGO 60637
THE UNIVERSITY OF CHICAGO PRESS, LTD., LONDON

82 81 80 79 78 77   9 8 7 6 5 4 3 2 1

Library of Congress Cataloging in
Publication Data

Woodward, David, 1942–
    The all-American map.

    "Published for the Hermon Dunlap
Smith Center for the History of Cartography at the Newberry Library."
    Bibliography: p.
    Includes index.
    1. Map printing. 2. Cerogra-
phy. I. Hermon Dunlap Smith
Center for the History of Cartography. II. Title.
GA150.W59   686.2'83
76–8099
ISBN   0–226–90725–2

for Ros

Contents

Figures

Figures

Preface and Acknowledgements

This book was written to document the history of an obscure printing process, wax engraving, that had a profound effect on the style of popular American maps and atlases for more than a hundred years, from the middle of the nineteenth century to the middle of the twentieth. The book is based largely on interviews and correspondence with the craftsmen engaged in the technique, and it is safe to say that without such documentation the technique and its practice would be largely forgotten twenty years hence. Wax engraving became obsolete in the 1960s, completely superseded by plastic drafting media, photocomposition, and offset lithography, all of which have undergone rapid improvement since the Second World War. Yet, while we are in the midst of this mid–twentieth-century revolution in cartography, it seems appropriate to pause and look back at a technique that reflected much of the character of the American graphic arts industry. Through its impact on the style of commercial map and atlas production, the technique is largely responsible for what may be regarded as the characteristically *American* map. By a detailed description of the effect of each aspect of the technique on cartographic style, the study attempts to make a methodological contribution in a little-known area of the history of cartography—the impact of production on product—and so adds a dimension beyond the documentary aims of the book.

The research on the process started with a single question,

which arose while I was perusing one of the better examples of American atlas cartography, the *Century Atlas of the World*, during the fall of 1965. At first sight the atlas looked remarkably modern, a characteristic that I was later able to ascribe to the exclusive use of letterpress type on the maps. But the atlas was dated 1897, some twenty years before type was to find widespread use in cartography. The question How was the type put on the map? led to a bibliographical search for material on an "unknown" American map reproduction process that employed letterpress type at the turn of the century.

As is often the case with a preliminary search for material on a problem in the history of cartography, Lloyd Brown's copious bibliography in *The Story of Maps* provided the starting point with a reference to an article by George Benedict entitled "Map Engraving" and dated 1912, fifteen years after the publication of the *Century Atlas*. "Map Engraving" turned out to be *wax engraving*.

Work by two other authors was also helpful in the early stages of the research. The late Erwin Raisz's *General Cartography* contained a brief mention of wax engraving, which prompted me to write to him for further information. Dr. Raisz then introduced me to Wallace B. Mitchell, who had been a wax engraver in the Boston area for many years. In addition, I found direct encouragement for the topic in a paper by Kenneth Nebenzahl entitled "A Stone Thrown at the Map Maker," published in 1961, a far-sighted paper which made a specific call for more studies in nineteenth-century American cartographic methods and contained one of the very few accounts of the basic steps in the wax-engraving technique.

Using the work of Benedict, Nebenzahl, and Raisz as a base, I searched especially through American technical printing journals for further evidence of wax engraving, as well as in other sources in the commercial graphic arts. Further study of the available literature, a close examination of a considerable number of maps printed by the technique, and a number of interviews with wax engravers and map publishers contributed to the present work.

In addition to the authors who provided the initial material for this book, I would like to acknowledge the help of many individuals and organizations. Particularly I would like to thank the following persons who contributed to interviews: J. Stanley Haas (Haas Wax-Engraving Company, Buffalo, N.Y.); Stanley Larson (Magnuson and Vincent Co., Cambridge, Mass.); Wallace B. Mitchell (Wallace B. Mitchell Co., Cambridge, Mass.); Ferdinand von Schwedler (Denville, N.J.); and Russell Voisin, who conducted an interview at Rand McNally and Co., Chicago, Ill., in November 1965 with staff members Edmund Brown, Walter Cahill, Art Klohr, and Minerva Lilja. For gently correcting some of my blunders on the early history of electrometallurgy as it related to wax engraving, I am indebted to the advice of Cyril Stanley Smith (Professor Emeritus, Massachusetts Institute of Technology).

The following individuals contributed information and views through correspondence and brief meetings: Crawford C. Anderson (Past Superintendent, Matthews-Northrup Division of the J. W. Clement Co.); C. A. Burkhart (Past President, A. J. Nystrom and Co.); R. E. Dahlberg (Professor of Geography, Northern Illinois University); L. B. Douthit (President and General Manager, George F. Cram Co.); Duncan Fitchet (R. R. Donnelley and Sons); Herman Friis (Director of Polar Archives, National Archives, Washington, D.C.); O. E. Geppert (President, Denoyer-Geppert Company); Beulah Hagen (Assistant to Cass Canfield, Harper and Row, New York); Elizabeth M. Harris (Division of Graphic Arts, Smithsonian Institution); Richard Edes Harrison (New York); Andrew McNally III (Rand McNally and Co.); Walter W. Ristow (Chief, Geography and Map Division, the Library of Congress); Norman J. W. Thrower (Professor of Geography, University of California, Los Angeles); and James M. Wells (The Newberry Library).

Acknowledgment is also due to the staffs of the following libraries: The Newberry Library (Chicago, Ill.); St. Bride Printing Library (London); Library of the Royal Geographical Society (London); The British Library (London); the Library of Congress (Washington, D.C.); the libraries of the University of Wisconsin (Madison, Wisc.); Yale University Library (New Haven, Conn.); and the New York Public Library.

The following companies were in correspondence with the

author: American Map Company (New York); California Electrotype and Stereotype Co. (Los Angeles, Calif.); Empire Wax-Engraving Co. (New York); Hearne Brothers (Detroit, Mich.); Chas. Holl Wax-Engraving Co. (San Francisco, Calif.); Kansas City Central Electrotype Co. (Kansas City, Mo.); Knipp Wax-Engraving Co. (Chicago, Ill.); A. J. Nystrom and Co. (Chicago, Ill.); Poole Printing Company (Chicago, Ill.); Potomac Electrotype Co. (Washington, D.C.); Rapid Grip and Batten (Montreal, Quebec); and the Roger A. Reed Co. (Reading, Mass.).

The research for the book was supported in part by the Graduate School of the University of Wisconsin. Further time for reshaping the work was granted by The Newberry Library and a British Academy Exchange Fellowship in summer 1973. My thanks go to these institutions.

Finally, I want to thank three individuals from a personal standpoint: Arthur H. Robinson, who has been a constant stimulus to me during and since my studies at the University of Wisconsin; Lawrence W. Towner (President and Librarian, The Newberry Library), who has given me every encouragement and help; and my wife Rosalind, to whom the book is dedicated.

1 *Introduction*

Wax engraving was a simple and ingenious process by which a metal printing plate in relief could be produced from an engraved mold (fig. 1.1). The engraver applied a thin layer of wax to a smooth plate, usually made of copper. He then engraved lines and symbols in the wax through to the metal beneath, using special tools of varying thickness and shape. The lettering could either be hand engraved or produced by pressing metal type through the wax. Fine line tints were often engraved with ruling machines.

After the engraving, the spaces between the lines were built up with wax to give depth to the subsequent printing plate. This plate was then cast from the wax by the process of electrotyping, a technique analogous to electroplating. Graphite was dusted on the mold to render the surface electrically conductive. The mold was then placed in an electrotyping tank where a copper or nickel shell was electrochemically deposited. When thick enough, the shell was removed from the mold, backed up with type metal, mounted on wood, and used as a printing plate. (See fig. 1.2.) The result was a relief-printing plate that could be printed on letterpress printing machines, a property that accounted in large measure for the versatility of the process in the nineteenth and early twentieth centuries.

What were the forces and conditions that gave rise to this technique at this time in history? To answer this question, it will

1

PRINTING PLATE

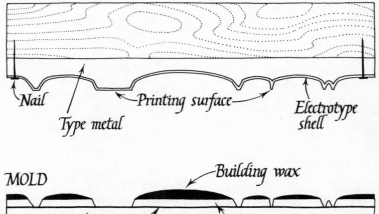

Fig. 1.1. The mold and printing plate: a summary diagram illustrating the wax-engraving technique

be useful not only to look at the immediate historical context of the process, which is discussed in the next chapter, but also to glance briefly at some of the major techniques of printing maps since the introduction of map printing in the fifteenth century. The origin, decline, and adaptation of some of these processes shed light on many of the same social and economic mechanisms that touched upon wax engraving.

In setting the process in its historical context, it is important to recognize that this study crosses what has been a rigid disciplinary boundary between the history of printing and the history of cartography. Yet, if the study is a hybrid, it is a natural hybrid, for cartography and printing have had a long and complex association in history, as Arthur H. Robinson has shown.[1] But since scholars in the history of printing and the history of cartography have rarely had an opportunity to glean each other's background, the remainder of this introduction will attempt to place wax engraving in the context of the history of

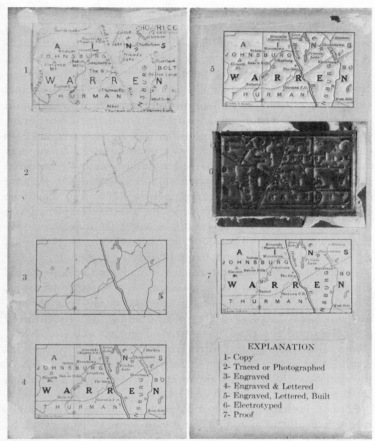

Fig. 1.2. Steps in the wax-engraving process. Courtesy Rand McNally and Co. Photo, Peter Weil

map printing, and I ask the patience of both cartographers and printing historians as they follow what to them may appear a well-trodden path.

THE HISTORICAL CONTEXT

For centuries the reproduction of graphic material has posed special problems for the printer. Until the nineteenth century, he tended to view these aspects as ancillary to his main job of reproducing manuscripts by means of movable type. As a result, the solution of these problems consistently lagged behind in improvements in the mainstream of printing technology throughout the first four centuries of its history. Thus the first printers who successfully faced the problem of printing scientific diagrams in the text did so some thirty years after Gutenberg had perfected a method of casting types. It was not until 1472 that a simple woodcut world map was printed by Günther Zainer at his press in Augsburg. When we consider that woodcutters had been cutting images in text on woodblocks for about seventy-five years before that date, it is strange that the map was not considered important enough to the dissemination of knowledge to warrant reproduction in multiple copies before 1472. Yet when we consider the stringent requirements the average map made on the common printing processes, it is easier to understand that this specialized form of graphic communication has been viewed as a special problem by the printer from the beginning.

Among the general characteristics of maps that require special attention by the printer are format, size, number of impressions, level of precision, degree of color, line and consistency in tone, ease of revision, reproducibility of fine detail, the frequent need for color, and the need to combine lettering and line work. When taken individually, such requirements are hardly exclusive to map work; but taken collectively, the characteristics of the map have always posed the unique combination of problems for the printer. It may be useful to enlarge upon these.

Format. Perhaps no other single medium of graphic communication is available in such a variety of formats. Maps may be printed on paper, cloth, cardboard; mounted on a large number of different materials, folded, flat, or rolled; and appear printed in books, newspapers, or atlases or as globes, wall maps, relief models, and so on. The choice of the format influences to a large extent the choice of printing process.

Size. Maps are usually larger than general illustrations and thus often require larger printing plates. The consequent bulk of these printing plates poses storage and handling problems for the printer.

Number of impressions. The variation in the number of prints required from the map printing plate is considerable, from two copies to a run of two million. Such a variation in demand was crucial in the emergence of a versatile map-printing process.

Precision. Ever since the introduction of scientific measurement into cartography in the sixteenth century, the matter of precision has also entered into the decision regarding the printing process. It is clear that if care and trouble are taken over precise survey measurements in the field, sufficient control should be exercised in the engraving and printing stages to maintain the same order of precision.

Consistency. As maps frequently appear in series or in single volumes such as atlases, it is extremely desirable for the printer to maintain consistency of color, line, and tone of a single map or from one map to another by even inking. Maps often exhibit large areas of flat tone which require extreme care by the printer in order to maintain evenness. In addition, in a map designed to consist of several sheets joined together, the importance of maintaining the "color" of the whole map is evident.

Ease of revision. A map is characteristically a reference document that incorporates rapidly changing data, so that the ability to revise printing plates has been a prime consideration in the choice of printing process.

Reproducibility of detail. Maps commonly include a large amount of fine detail that has required printing processes capable of reproducing extremely fine lettering, lines, and symbols.

Color. A map is a complex document often containing classes of information that must be distinguished easily from other classes of information without being given undue emphasis. Color has proved to be one useful means of achieving this end, and attempts have been made throughout the history of map printing to develop a process that could reproduce color economically.

Lettering. The problem of lettering is one of the most pervasive in the history of map printing. It is not so much the problem of the execution of the lettering itself, although this can present difficulties, as it is a problem of combining lettering and line work in close proximity. This places severe limitations on the use of movable type in map printing plates and has led to a number of ingenious experimental processes to overcome the problem.

Until the invention of lithography at the close of the eighteenth century, the choice of processes open to the map printer was simple: woodcut or copperplate engraving. The processes were very different in many respects, beginning with the structure of the printing surface itself. While the woodcut was a *relief* printing process, printing from the raised portions of the block, the copperplate technique was an *intaglio* process, in which the ink was transferred to the paper from furrows, points, or flicks engraved in the plate. The concept of the work was therefore different from the beginning. The copper engraver cut the plate where it would print; the untouched portions of the plate formed the nonimage areas. Conversely, the woodcutter cut away those portions of the block which were not intended to print and left the block untouched in the intended image areas. At the engraving stage, the two techniques were thus totally different, involving different tools, and requiring different skills of different artisans.

The differences extended to the presswork stages. In intaglio printing from copperplate engravings, the ink applied to the plate was wiped or rinsed from the nonimage surface and allowed to remain only in the incisions. The printing was done on rolling presses, where great pressure was used to force the paper somewhat into the ink-filled incisions so that the ink was drawn out and imprinted on the paper. In relief printing from woodcuts, the ink was applied to the uncut portions of the surface of the block, and the impression was drawn on a common hand press originally designed for movable type and using moderate vertical pressure.

For the average printer, the printing of woodcuts was a straightforward process. Impressions could be drawn easily with equipment already on hand. In addition, woodcut blocks and type could be printed together at one and the same time, so that the technique was well adapted to printing maps in books. Because of its simplicity and directness, the woodcut process was used during the early years of printing not only in Western

Europe but also wherever only simple equipment was available. Thus the earliest known map printed in what is now the United States was a woodcut printed on a hand press in the seventeenth century.[2]

Although no hard statistics are available, it appears that the woodcut was the more common technique for reproducing maps from 1472 until the middle of the sixteenth century. The technique had been particularly popular north of the Alps, in the Rhine Valley, Bavaria, and Switzerland, where a strong tradition of woodcarving existed.[3]

The woodblock was a resistant medium for the craftsman in the sense that it was difficult to work. Fine lines were hard to fashion in the close-grained wood, and lettering was particularly troublesome, which led to the practice of inserting letterpress types or small stereotype plates into the block. The latter practice was introduced around 1530 and became a common refinement to the woodcut technique. Even with this improvement, the appearance of woodcut maps is characteristically rough and simple, reflecting the resistance of the medium.

The rise of the Italian and later the Flemish engraving workshops of the sixteenth century heralded an era of intaglio engraving for maps which was to last three centuries. Copperplate engraving had existed from the beginning of map printing: the 1477 edition of Ptolemy's *Geographia* was printed from copperplates, and the 1478 Rome edition of the same work is a masterful example of engraving. Yet the technique did not assert its dominance until the middle of the sixteenth century. Among the disadvantages were the sheer labor of inking the plates and the special intaglio rolling presses that had to be developed for printing them. In addition, the copperplate technique was not well suited to printing maps in the text, since to accomplish this, two printings were necessary. Thus, most of the copperplate maps appearing in books were tipped in on separate sheets, a procedure that added extra cost to an already expensive process.

Nevertheless, the advantages were real. The technique afforded considerable freedom to the engraver; with a simple tool, the burin or graver, he could fashion both detailed line work and lettering with a grace that the woodcut had never attained. The plates could be as large as press and paper would allow. Revision of the plates was simple compared to the woodcut technique: plates could be reworked and deepened, and sometimes lasted for centuries. The copperplate technique held its own longer than any other single map printing process, from the middle of the sixteenth to the middle of the nineteenth century.

With the invention of lithography by Alois Senefelder around 1796, the choice of a printing process for maps was no longer limited to the woodcut and the copperplate engraving. Lithography created a turning point in the history of map printing, upsetting the neat distinction between relief and intaglio techniques. At this point it is no longer particularly helpful to draw the basic distinction among printing processes according to the nature of the printing surface. With lithography came the ability to transfer images directly to the lithographic stone, no matter how they were printed originally. A copper engraving was frequently transferred to the lithographic stone, from which it was then printed. Woodcuts, type, pen drawings, etchings, aquatints, mezzotints, and so forth, could all be transferred, at first directly by what was called the anastatic process. When photography appeared in 1839, lithography was able to take advantage of the new art, and photolithography was a reality by 1860. In addition, the lithographic stone was subject to different treatment: it could be etched or engraved as well as simply drawn upon with greasy ink, so that the use of the term *planographic* (or printing from a flat surface) to distinguish lithography from relief or intaglio techniques is misleading.

By 1850, lithography had broken the monopoly of copperplate engraving that had lasted for three centuries. It was an extremely versatile autographic medium and found favor with cartographic draftsmen, who could draw with pen and ink on paper and transfer the drawing to the stone or draw directly on the stone surface, and who could render a wide variety of tones with crayons and other grease-based media. Yet both copper engraving and lithography were difficult to adapt to the growing applications of mechanical power to printing in the nineteenth century. The steam press of König (1812) was designed expressly for relief-printing surfaces such as foundry type and wood engravings, and could not accommodate the engraved copper-plate or the lithographic stone. It was not until 1851 that a successful steam cylinder-press was designed for lithographic use by George Sigl, producing six to eight hundred copies an hour, or approximately eight times the output of the hand lithographic press. Yet the cylinder action, in which the paper was placed between the printing plate and the cylinder, was to be replaced by the rotary principle, in which the printing plate was curved around the cylinder. A major requirement for rotary presses in lithography was the flexible printing plate, and therefore zinc gradually replaced the rigid and bulkier lithographic stone.

Offset lithography revolutionized the printing of maps. In offset printing, the ink is transferred from the plate to an intermediary surface which then imprints, or "offsets," the image on the paper. While the offset principle had been used in printing on tinplate since the 1870s, it was not until 1904 that Ira W. Rubel successfully developed the offset press for printing on paper. In Rubel's press, impressions were transferred to a rubber blanket before printing, thus reducing wear on the printing plates and enabling various textures of paper to be used.

The market for cheap illustrated books, magazines, encyclo-pedias, and atlases, generated by the newly literate masses of the mid-nineteenth century, could not wait for these improvements in the application of power to lithography. It was thus left to a relief process, for which the early steam presses had been designed, to fulfill this need. The wood engraving, primitive though it was, met the need until the commercial introduction of line blocks in the 1870s and halftone blocks in the 1880s. By 1900, the wood engraving as a commercial reproduction me-dium was all but dead.

The wood engravers reached a high degree of technical virtuosity in illustration work, and the woodblocks were durable in the power press, but for maps the technique proved to be clumsy. In particular, the printing of large maps with a great deal of lettering created serious problems. Printers sought to overcome these problems by developing cumbersome machines to cut slots through the wood to receive letterpress type for the lettering.

Chapter One

There were many experimental attempts in the 1820s and 30s to develop a relief-printing process for illustrations to replace the wood engraving; the patent records abound with them. Of these attempts, wax engraving ranks as one of the successful experiments, particularly in the United States, where it achieved widespread commercial use.

2 *Origins and Development*

Wax engraving was one of a number of experimental printing processes in the nineteenth century intended for commercial use in large editions on a power press. As a relief process, it was intended to replace wood engraving, which nevertheless held its own in certain classes of graphic reproduction until the end of the nineteenth century. The boxwood used by the engravers was a durable material from the pressman's viewpoint, and hundreds of thousands of impressions could be taken from such blocks. But boxwood was hard to work for the engraver, especially for a map engraver faced with hundreds of small names.

The task was to find a medium soft enough to engrave, yet which could produce a printing plate that was sufficiently resistant to wear in the press. The dilemma was that the ideal durable material for printing was metal, yet the materials comprising the traditional molds in which metal could be cast (sand, plaster of paris, and so forth) were difficult to work.

By about 1830, there were three possible ways of producing a metal relief-printing plate: metalcut, relief etching, and stereotyping. The metalcut, in which the printing surface was incised directly into the metal, obviously offered no improvements over the woodcut from the engraver's viewpoint, and while it is difficult to tell one from the other from the impressions alone, it is inconceivable that the technique was ever used extensively for maps. Relief etching, developed and used by William Blake in the early 1800s, was an ingenious technique but had distinct

11

Chapter Two

disadvantages as a commercial process. By drawing or painting with varnish on a copperplate and etching the remainder of the plate with acid, Blake was able to produce a shallow relief-printing surface. Yet it was obviously too shallow to withstand mechanical inking and presswork, and it was not until the 1850s that Charles Gillot was able to produce the first commercially feasible relief-etched block.

Stereotyping, which found its main use in the "freezing" of whole pages of movable type for reprinting purposes, was also used to duplicate graphic printing plates as early as the fifteenth century. Molds were taken in fine sand, plaster of paris, papier-mâché (flong), and a number of other materials that could withstand the heat of molten type metal without disintegrating. None of these materials, of course, was suitable for engraving. One solution to the problem came with the so-called "clay process" in which fine clay was used for the mold in stereotyping instead of the usual gypsum or papier-mâché. The clay lent itself to original engraving, and was apparently used by the New York firm of Fisk and Russell in the 1850s.

The breakthrough occurred with the commercial application of electrochemical deposition. At last a metal plate could be cast from an easily engraved soft molding material without destroying the mold. Now the plate could be built up slowly on the mold, molecule by molecule, with cold metal. The phenomenon of electrodeposition had been observed earlier, notably by William Cruikshank about 1803–4, and Michael Faraday had made the first exact quantitative study based on experiments conducted between 1830 and 1833, which led to a satisfactory theory of electrochemical deposition.[1]

The advantages of electroplating were immediately recognized in the printing world and quickly adapted to the duplication of printing surfaces in a technique known as electrotyping. As with all such discoveries in the history of printing, there are several contenders for the honor of being recorded as the inventor of electrotyping. Rather than to become entangled in arguments about who did what first, it is probably wiser to consider these claims as independent and contemporary.

There were two main obstacles to the commercial development of electrotyping, both of which had been more or less overcome by the early 1840s. First, there was a problem in finding a battery powerful enough to deposit the metal electrochemically in a reasonable length of time. The Daniell battery (1836) was successively improved upon by the Grove battery (1839) and the Smee battery (1840).[2] The last used platinum or silver plates instead of copper to avoid the adherence of hydrogen bubbles on them (polarization) which reduced the effective surface area of the electrodes. Second, there was difficulty in finding an effective conducting material which could be applied to the surface of the mold. In 1840, a Mr. Murray was awarded a silver medal by the Royal Institution for discovering that wax could be made electrically conductive by brushing it with graphite (otherwise called plumbago or black lead, both misnomers, as the material contains no lead).[3]

An announcement of practical electrotyping came from M. H. von Jacobi in St. Petersburg on 5 October 1838, and it was soon followed by independent claims of invention by two Englishmen, C. J. Jordan and Thomas Spencer. The latter has provided us with an early account of a form of wax engraving:

> My first essay was with a piece of thin copper plate, having about four inches of superficies, with an equal sized piece of zinc, connected together with a piece of copper wire. I gave the copper a coating of soft cement, consisting of beeswax, resin and red earth—Indian or Calcutta red. The cement was compounded after the manner recommended by Dr. Faraday in his work on Chemical Manipulation, but with a larger proportion of wax. The plate received its coating while hot. On cooling I scratched the initials of my own name rudely on the plate, taking special care that the copper might be thoroughly exposed. This was put into action in a cylindrical glass vessel about half-filled with a saturated solution of sulphate of copper. . . .
>
> It was then suffered to remain, and in a few hours, I perceived that action had commenced, and that the portion of the copper rendered bare by the scratches was coated with a pure, bright, deposited metal, whilst all the surrounding portions were not at all acted upon.
>
> I now saw my former observations realized; but whether the deposition so formed would retain its hold on the plate, and whether it would be of sufficient solidity or strength to

bear working, if applied to a useful purpose, became questions which I now endeavoured to solve by experiment. It also became a question whether, should I be successful in these two points, I should be able to produce lines sufficiently in relief to print from. The latter appeared to depend entirely on the nature of the cement or etching ground I might use.[4]

In April 1839, an announcement appeared in the *Athenaeum* that Professor Jacobi of St. Petersburg had "found a method . . . of converting any line, however fine, engraved on copper, into a relief, by a galvanic process." On publication of the announcement in the *Mechanics Magazine* for May 11, one C. J. Jordan was prompted to write that he had been conducting similar experiments as early as the summer of 1838.[5]

In America, Joseph A. Adams had been working in 1839 with the process of electrotyping, specifically for producing duplicate plates from wood engravings, the first of which appeared in *Mapes's Magazine* in 1841.[6] As the quality of extremely fine lines could be maintained, the process was a great improvement on the previous duplication of wood engravings by stereotyping, and the technique became common in the last half of the nineteenth century.

It was on the invention of electrotyping that the development of wax engraving depended. It was no great mental jump from the idea of duplicating existing printing surfaces in wax to engraving original work in the same medium and to use the same process of electrotyping in the manufacture of relief-printing plates. We find a number of descriptions of essentially the same process during the late 1830s and early 1840s. Some were patented but apparently rarely used, such as the unnamed process of William Tudor Mabley.[7] Others were used but apparently not patented.

The work of two men during those early years stands out as worthy of specific mention, since their efforts, although not commercially successful, were used enough to be noticed, and laid the groundwork for later developments of the technique. In Britain, Edward Palmer was the pioneer, while in the United States the efforts of Sidney Edwards Morse claim special attention.

Fig. 2.1. Morse family portrait painted by Samuel F. B. Morse around 1810. From left to right, Elizabeth Ann, Sidney Edwards, Jedidiah, Samuel Finley Breese, and Richard Cary. Courtesy The Smithsonian Institution

Chapter Two

SIDNEY EDWARDS MORSE
(1794–1871)

Son of a famous geographer, Jedidiah Morse (1761–1826), and brother of the celebrated inventor Samuel Finley Breese Morse, Sidney E. Morse was well steeped in an environment of letters and science. Much of his energy was consumed in editing and journalism. By 1812–13, at the age of eighteen, he was writing for the *Columbian Sentinel*, and in 1815 he became the first editor of the *Boston Recorder*, the first religious newspaper in the United States, published by Nathaniel Willis. In 1823 he founded, with his younger brother Richard Cary Morse, the *New York Observer*, a religious weekly of which he remained the senior editor until 1858.

In geographical writing and editing, Sidney Edwards Morse was admirably placed for a promising career. His father's name had already become a household word in geography. As Ralph Brown puts it in his fine biographical article on Jedidiah Morse, "the vast majority of homes possessed a well-thumbed Morse . . . by 1800, his name was familiar to the reading public [as] the father of American geography."[8] As Jedidiah aged, Sidney took over more and more of the editing of his father's geographical works, and after the latter's death in 1826 continued to bring out new editions of the textbooks, gazetteers, and atlases.

Sidney E. Morse's inventive talents were made manifest in many ways, from the patenting of a flexible piston pump with his brother Richard in 1817 to the invention and improvement of a bathometer for deep-sea soundings during his later years. But the invention for which he will be known best is his cerography, or wax engraving, which he was working on as early as 1834 with his engraver, Henry A. Munson. It was through the medium of the *New York Observer* that cerography was announced, giving us a glimpse of an otherwise secret process. The first announcement came on 29 June 1839, with a map of Connecticut (fig. 2.2), and its importance in the origin of wax engraving warrants its reproduction in full:

> MAPS-NEW MODE OF ENGRAVING. —Several years since, as most of our readers will remember, we commenced publishing in our paper, wood-cut maps of the different states of the

Fig. 2.2. Sidney E. Morse's map of Connecticut, appearing in *New York Observer*, vol. 17, 1839

Union, and of other countries, with the intention of furnishing in the end, a complete Atlas, suited to the use, especially, of religious-newspaper readers. When the plates of the U.S. maps were nearly finished, a part of them were consumed by the fire in Ann Street, and before this loss was fully repaired, they were again injured in the great fire. When, at length, they were all done and ready to be blocked, we became dissatisfied with the style of the engraving. They were, indeed, very good wood-cut maps; but from the nature of wood-cut engraving, we were under the necessity of omitting roads, names of towns, and other information, to such an extent, that when we compared them with the copper-plate maps from which we copied, they lost in our eyes nearly all value. In reflecting on the matter, we became satisfied that a new mode of engraving was practicable, by which map-plates could be easily made, containing all information on copper-plate maps, and yet printed, in connection with type, under letter press. Accordingly, we commenced our experiments, and persevered, until they were crowned with complete success. The map of Connecticut, which we give on our last page, is from a plate obtained by the new method. Wood cut engravers, or persons conversant with their art, will see at once that the information on this map could not be given in relief on wood, except at an expense which must deter any on[e] from attempting it. By the new method it is very rapidly done.

As the inventor of a new art, we shall be allowed, we suppose, the privilege of giving it a name. We accordingly name it, —Cerography. In a few weeks we hope to be able to prepare a specimen sheet, which will show, that for maps, music, and some other kinds of engravings, Cerography, with proper attention to the press-work is capable of furnishing prints that will make a very near approach in beauty to those from copper-plates. If this can be done, the superior facility of the engraving, the durability of the plate, and the rapidity of the printing will give it great advantages. We need not inform persons acquainted with newspaper printing under a Napier press, that we do not refer to the map in this paper as a specimen of any thing but the amount of information which can be given by the new art. No judgment can be formed of it by the delicacy, or beauty, of which the style is susceptible.

We shall now go on with the preparation of the plates for the Atlas, and trust that our readers will find in the superior value of the work a compensation for the long delay in its appearanoce [*sic*]. Experiments have cost us several thousand dollars; but when we consider the facilities which they have given us for furnishing valuable information to the readers of the Observer, we feel that in this way alone we shall be amply repaid.[9]

Three main points emerge from this statement. First, we are made well aware that Morse and his associate found wood engraving, which they had used in the *New York Observer* for several years, unsatisfactory because of the crudeness of the technique. Second, the extract reveals a reverence on the part of the author for the delicacy of copper engraving, which was the object of their emulation. Finally, the whole announcement is an apologia: for the length of time the new process had taken to develop, and for the quality of the engraving which, it is pointed out, cannot represent the potential of the new art.

In addition to providing relief-printing plates for maps in the *New York Observer*, Morse offered atlas supplements as an inducement to subscribers. One of these, the *Cerographic Atlas of the United States*, was issued in parts between 1842 and 1845. Other publications were the *Cerographic Bible Atlas* (1844) and the *Cerographic Missionary Atlas* (1848). The scheme had been announced in the *Observer* in 1841, and it appears that the promises were fulfilled:

Our plan of cerographic maps embraces—
  1. A complete Atlas of the United States, in about twenty maps.
  2. An atlas of the various missionary stations throughout the world.
  3. A complete Bible Atlas

Each of these atlases, when completed, will be done up in a cover, and sent gratuitously to such of our subscribers as are entitled to them, and to all who become subscribers during the first three months of the present year, and pay for the term of two years.[10]

Morse's main commercial use of his cerography was in association with the publishing firm of Harper and Brothers, which undertook to publish the *North American Atlas*, Morse's *School Geography*, and a large wall map of the United States. The four existing contracts relating to these enterprises, reproduced in the Appendix (pp. 129–38), shed a great deal of light on the map printing and publishing practices of the time. Of the three projects, the *School Geography* seems to have been the most lucrative. This was an illustrated quarto which retailed for fifty cents. The first edition, published in 1844 in a run of 100,000, had sold 73,000 by October 1845, and Harper's were considering a reprint of 20,000.[11] There were new editions in 1847, 1851, and 1853, in which year the market had dropped to 70,000. A fire at Harper's during 1853, which destroyed many of the cerographic plates, combined with the rapid crumbling of the reputation of Morse geographies, made further editions uneconomic.

The *North American Atlas* was less successful than the *School Geography*. In 1845, of 5,000 copies printed, only 1,000 had been sold, but it appears that since the atlas was issued in parts, many potential buyers were waiting for the whole atlas to appear before buying.[12] The record does not reveal how many parts were published or how many copies were ultimately sold.

The third project which Harper and Brothers undertook for Morse was *Harper's Map of the United States and Canada*, a wall map mounted on rollers published in 1847. Four versions of the map were available, costing from $2.00 to $2.50, depending on the number of colors used.[13]

The cerographic maps of Sidney Edwards Morse were not merely straight copies of his earlier work in the 1820s or of his father's maps illustrating the geographies and gazetteers. We have evidence that Morse's geographer, Samuel Breese, took pains to obtain the most accurate and up-to-date information available. In a copy of the *Cerographic Atlas of the United States*, once belonging to Increase A. Lapham, now in the collections of the State Historical Society of Wisconsin, there is a manuscript note: "If you have any more new country or towns will you be so kind as to send them to me—When we print some on better paper I will send a complete and perfect copy—B[reese?]."

The cerographic process, or wax engraving, was Morse's main claim to originality, and one of which he was proud, as the written record shows. Yet he was always stressing the need for secrecy of the method, as is revealed in a letter to his brother Richard in 1845: "Be careful, and not leave any letter [*sic*] to you where they will be read by others, especially my letters on business. S.E.M."[14] While Morse and his chief engraver, Henry Munson, had been working on cerography since 1834, it was not until 1848 that he patented the technique. By that time, the art was well known in the trade on both sides of the Atlantic.

The late date of Morse's patent makes it difficult to reconstruct the earlier stages of his technique. There is no hard evidence that Sidney Edwards Morse used electrotyping in the early stages (that is, pre-1840). Two theories emerge, both based only on supposition and circumstantial evidence. On the one hand, many European scientists had been experimenting with electrometallurgy (though with no practical application in mind), and Faraday's ionic theory was well known by 1834. News of these experiments could have been transmitted by Morse's brother Samuel Finley Breese Morse, who on 21 September 1832 wrote that he was visiting old friends and renewing old associations in London.[15]

But on the other hand, there are several reasons why Sidney Edwards Morse would not have developed the art of electrotyping and applied it to his cerography before all the contenders who crowd the literature with their claims in 1839–40. In the first place, the fact that Sidney was the brother of a man the stature of Samuel is absolutely no guarantee that he shared the latter's knowledge of the new possibilities of electrotyping suggested by Faraday's experiments. The development of electrotyping as a commercial process was not contingent on the theoretical work of Faraday, even though that theory helped refine the technique later in its development. And the supposition that Samuel Finley Breese Morse, on his return from England (just off the boat as it were), should brief his brother on such matters is unwarranted.

The possibility that Sidney Edwards Morse (who had considerable inventive talent) had discovered electrotyping before

Jacobi, Jordan, and Spencer is intriguing, but should be discarded. His newspaper reports with their boasting of the mysterious use of wax reveals enough of his character to make us conclude that if his invention had involved the use of electrodeposition from the beginning, he would have used the magic of electricity in christening his invention and called it something like *electrocerography*.

I prefer a theory of Professor Cyril Stanley Smith who surmises that the lines and symbols cut and impressed in a layer of hard wax on a flat metal plate were copied by casting a fusible metal against it to give a relief-printing surface directly. Several low-melting alloys were known in Morse's time, notably Wood's metal, an alloy of bismuth, lead, cadmium, and tin (5 Bi, 3 Pb, 1 Cd, 1 Sn) which melts at about 70° C., well below the boiling point of water. The alloy used for backing the electrotype in Morse's patent of 1848 is similar to this with the omission of cadmium, which would melt at about 95° C. A very hard wax or a mixture of wax and resin would withstand such temperatures, and the cast metal plate would wear at least as well as the softer composition of type metal.

Morse could also have produced an inverse replica of his wax-coated surfaces by using one of the techniques of *nature printing*, which for about a century had had some success on both sides of the Atlantic for the copying of soft plant specimens in the direct illustration of books on botany.[16] Although this sometimes involved making an impression of the original, backed by a steel plate, in a lead sheet, a common method was also to make a mold of plaster or other material against which the metal was subsequently cast as in stereotyping. With Morse's intaglio original, two mold inversions would be necessary in order to obtain the details in relief for printing. There may have been some echo in his mind of the traditional lost-wax process (*cire-perdue*), long used in casting metal art objects with fine surface detail. On the whole, however, the direct casting of fusible metal against a hard wax surface seems to be the most plausible beginning for Morse's process that was later modified by the incorporation of electrotyping, as described in his patent.

It thus seems reasonable to suppose that Morse's patent of

1848, with its clear addition of electrotyping in the process of cerography, represents an improvement over his original technique as conceived as early as 1834, but cannot be construed as representing it in every detail. Morse thus takes his place with the other users of electrotyping from 1839 onward to produce a relief-printing plate. His technique differed from that of his competitors (especially Palmer) in the amount of electrodeposition during electrotyping. His patent clearly states that, after engraving in the wax compound, the plate is immersed in a solution of copper sulphate and connected to a battery, "so that copper may be deposited in all the lines and marks of the writing or drawing by the well-known electrotype process." The copper was to be deposited only in the engraved lines; when the copper began to be deposited elsewhere on the mold, it was to be taken out of the electrolyte and backed with a molten alloy (tin, bismuth, and lead) to form a bimetallic printing plate. The alloy, forming the nonprinting surface, was then lightly etched to lower it, a difficult process in that the plate had to be removed from the acid just before the copper was attacked. By that time, the alloy had been sufficiently broken down to be able to be removed with a stiff brush, thus forming the hollows or lower parts of the printing plate.

The contribution of Sidney Morse was important in cartography in that it unquestionably demonstrated the use of wax as a molding medium to produce a relief-printing surface of sufficient quality for maps; hence the term cerography. There is also no question about his taking advantage of electrotyping when it was developed enough for him to do so, and at least as early as 1848, the date of his patent. But we are unsure of the time when he applied the art of electrotyping to his process, and we are likely to remain so until more hard evidence appears.

The Development of Wax
Engraving in the United States

Wax engraving found much wider use in the United States than in any other country. The several reasons for this will be examined in detail later in this chapter. In a nutshell, the technique appeared on the scene in time to serve the needs of the rapidly growing mass market for popular maps in guidebooks, encyclopedias, school and family atlases, railroad timetables, and so on. Using figure 2.3, we can recognize four main periods in the development of wax engraving in the United States: (1) an experimental period from 1840–50 represented largely by the work of Sidney Edwards Morse; (2) the gradual acceptance of the technique by a few firms from 1850–70; (3) the major period of the technique, from 1870–1930, when the vast majority of American commercial maps and atlases employed the technique; and (4) the period of replacement by other techniques, specifically offset lithography, from about 1930–50. Very few maps indeed were wax engraved after 1950.

During the entire 110-year life span of wax engraving, from about 1840 to about 1950, nearly three-quarters of regional, national, and world atlases commercially published in the United States used the technique. During its peak between 1910 and 1920, the figure rises to 95 percent. While an actual sample count of separate maps has not been made, we may expect that the figure would hold good for maps as well as atlases.

The Experimental Period,
1840–50

We have already traced the origin of the technique in the United States. There were several independent developments elsewhere that occurred about the same time: Edward Palmer's glyphography in England and Volkmer Ahner's process in Austria were examples. A study of the various claims to primacy in the invention of the technique is not particularly rewarding: the processes were all developed around 1840 and were based on the invention of electrotyping.

As we have seen, the pioneer in the United States was Sidney Edwards Morse. By 1845 he had sold most of his plates to Harper and Brothers, and the lack of map-related subjects in his correspondence after that date suggests that he was more generally interested in pursuing his editorial activities for the *New York Observer* and proslavery writings, although in 1860 he republished his *Cerographic Bible Atlas,* using plates that had been copyrighted in 1843. In addition, Harper and Brothers

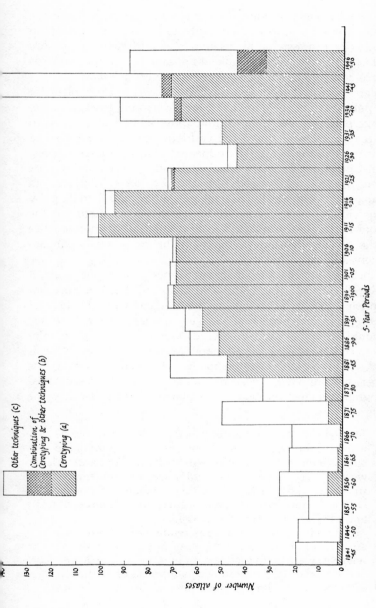

Fig. 2.3. Graph illustrating the numerical significance of wax engraving in American commercial cartography. Statistics based on a count of 1232 regional, national, and world atlases printed in the United States and preserved in the Library of Congress

continued to publish editions of his *School Geography* in the 1850s. There is, however, no evidence of any active role on Morse's part to develop his technique of cerography to a more sophisticated level. For example, it appears that Sidney Edwards Morse never used letterpress type on any of his maps.

Maps in school geography texts posed a special problem for the publisher and printer that Morse was able to help solve. Copperplate engraving and lithography, while better suited to the reproduction of fine detail and lettering, could be used to print maps in the text only by means of an additional impression. Consequently, maps were frequently printed on separate sheets, folded, and tipped into the text on thin paper, an arrangement not ideally designed to withstand handling by schoolchildren. An alternative was to have the maps for the school geographies printed in a separate atlas, a common practice in the nineteenth century, but nevertheless somewhat inconvenient and expensive.

Since the wax-engraved map was printed from a relief plate, it was well suited to use in the textbook, and it is therefore not surprising that if one man can be credited with the idea of starting to include maps in school geographies on a regular basis, it would have to be Sidney Edwards Morse. His first geography text to contain wax-engraved maps was the twenty-seventh edition of his *School Geography*, which Jedidiah Morse, his father, had initiated. To accommodate the fifty-three detailed cerographic maps, the book was quarto. Exman describes the volume thus:

> This 9-by-12 inch size, established by Morse because he wanted detailed maps, made the *School Geography* the largest book that generations of American boys and girls were to carry home from school, always a problem, for smaller books were forever falling out when strapped in with it. In addition to the maps, there were 72 pages of text and 143 [pictorial] wood engravings.
>
> Bound in "muslin," the book retailed for 50 cents. On an order of one thousand copies or more, the brothers (Harper) were to give a discount of not less than 30 per cent. Did school district or statewide adoptions of textbooks account for such large group sales? At any rate, the new

"cerographic" process for producing the maps brought costs down to a point where the brothers could claim that their edition of Morse's *Geography* was priced lower than any competing book.[17]

The number of maps in American school geographies continued to increase. In a study by Sahli, twenty-five of the forty-nine geography texts published before 1840 contained maps, while only ten of the 1840–90 texts analyzed by Culler had *no* maps.[18] Part of this increase was no doubt facilitated by the use of wax engraving.

The Transitional Period, 1850–70

The Morse name, made famous by Jedidiah Morse and his son Sidney, was perpetuated further by one Charles W. Morse, who published several maps and atlases during the 1850s, of which the *Diamond Atlas* (1852) is his best-known work. While no family connections between Sidney and Charles have yet been established, despite an examination of the genealogical records, Charles was clearly not averse to having his atlases and maps considered as being in the true Morse tradition. Consequently, he frequently made references to the "cerography" by which his maps were produced, and the titles of his maps characteristically begin "Morse's Cerographic Map of. . . . "[19]

Charles W. Morse was associated with Rufus Blanchard, a Chicago publisher, having some of his maps published by the latter. Blanchard published numerous city guides containing wax-engraved maps and several separately published maps, mainly of Chicago and surrounding cities and states. He was joined by his nephew, George F. Cram, who had spent three years in the Union Army. At Blanchard's death in 1867, Cram took over the business, which was renamed the George F. Cram Company, still in existence today.[20]

Another center of wax engraving during this transitional period was Buffalo, New York. The firm of Jewett and Chandler was using the technique in the 1850s for illustrations in the reports of the United States Patent Office, and later for maps. Several engravers at Jewett and Chandler took their skill with them either to other companies of their own founding, or to companies new in the map-printing business that sought technical information. An engraver from Jewett and Chandler, for example, was sent to the firm of Fisk and Russell in New York City to demonstrate the process.[21] Also, Charles H. Waite introduced the techniques to Rand McNally before 1872.

In the minds of some commercial engravers, wax engraving began in Buffalo during this period, and none of the engravers I interviewed had ever heard of Sidney Edwards Morse. There is no documented connection between Morse's early work and the main period of wax engraving. According to William Haas, an apprentice of Henry Chandler at Jewett and Chandler, "The Wax Process had its beginning in the year 1855 being invented and perfected by Mr. Henry Chandler of Buffalo, New York . . . . then he thought of the plan for stamping type into the waxed plate, first using single letters, then words and several words together."[22] Haas's assertion as to the inventor may be questioned, but there is certainly no mention of the use of type in wax engraving in any patent specification before that of John McElheran, in which is described an incredibly cumbersome but ingenious machine for impressing individual letters into wax. In a sense, it was also an early composing machine.

This transitional period illustrates the great interdependence among the few companies that were practicing wax engraving and the strong family-oriented nature of the firms. (See fig. 2.4.) There were few well-documented technical innovations, although the use of type in wax engraving seems to have arisen largely in this period, at least in the United States, and this was to have a profound effect on the visual character of American commercial maps. The impact of photography was not great; wax engraving was a nonphotographic medium, and the only application of photography would have been in the transfer of the compilation to the surface of the wax. There is no evidence that this was done as early as the 1850s or 60s.

The Main Period, 1870–1930

During the experimental and transitional periods, the use of wax engraving was sporadic and uneven, and the large majority of maps and atlases were lithographed or engraved intaglio on

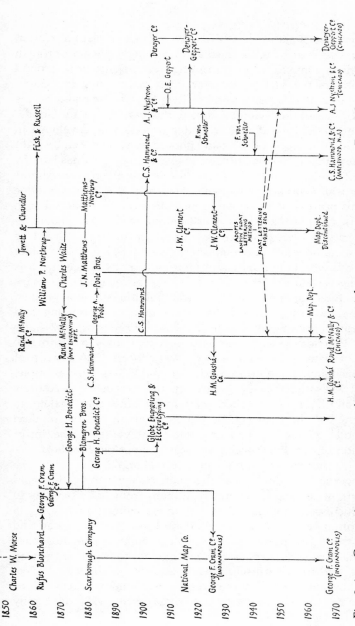

Fig. 2.4. Connections among some American commercial map-printing companies. From Special Libraries Association, Geography and Map Division, *Bulletin* 79 (1970): 3

copper or steel plates. The 1870s saw the beginning of a trend in which wax engraving assumed an increasingly dominant role in American commercial map printing, reaching its peak around the time of the First World War, then maintaining this dominance until the 1930s. Several writers have commented on the significance of the technique during this period. J. Paul Goode, writing in 1927, said that "a better process was developed in America after electro-typing and photography had been brought into use, that is the wax process of engraving. . . .̇ It has had the largest influence on map making in this country (U.S.). All our better wall maps and school geography and atlas maps are done by this process. It has done more than any other one thing in America to put the map to work among millions of people."[23] Sherman, in 1915, has the following comment: "While it is true that a number of map publishers have supplanted the older methods with lithography and photo-engraving processes, it is equally certain that the majority of the better grades of maps issued today by the largest American publishers are the product of a combined process of wax-engraving, electrotyping and letterpress printing."[24] Again, the *Inland Printer* of 1922 states that "Practically every map used in school study work for the past 25 years has been made by the wax process."[25] A measure of the technique's importance to map publishers (for example, Rand McNally) may be gauged by the statement of Andrew McNally III in 1956: "There were many other map makers in America at the time (1872), but none employed the modern methods of making map engravings in wax. . . . The introduction of this single technique was responsible for the Company's instantaneous success in map making."[26]

There is little doubt that the main reason behind the rapid growth in demand for commercial maps from 1870–1930 lay in the increased mobility of a rapidly growing population (38 million in 1870; 123 million in 1930) through the railroad and later the highway network. A mobile population demands maps for many reasons: to plan where it is going, to help find the way there, and to orient itself when it gets there. There is the less obvious but nevertheless important need for maps to remind one of home: the town view or state map on the wall, the family atlas in the bookshelf.

It is not surprising that the following announcement appeared in the October 1872 issue of Rand McNally's *Railway Guide:*

RAND MCNALLY & CO.

PRINTERS

ENGRAVERS, ELECTROTYPERS AND MANUFACTURING STATIONERS.

CONSECUTIVELY NUMBERED RAILROAD TICKETS

Map Engraving a Specialty.
All Kinds of
RELIEF LINE ENGRAVING
Promptly Executed.
(Etc.)

This advertisement marked the foundation of Rand McNally's Map Engraving Department, based on the technique introduced in that year to the company by Charles H. Waite, formerly an engraver with the Jewett and Chandler firm in Buffalo. The "Relief Line Engraving," which is none other than wax engraving, had a profound effect on the costs of map production. In the words of James McNally:

> In 1872, the Company introduced the then new "Relief
> Line Engraving" in the making of maps—a process which
> entirely revolutionized both the method and the cost of
> production, and reduced the price to purchasers by several
> hundred percent. A map that can now be bought for
> twenty-five cents, or a dollar, used to cost, under the more
> expensive ways of production, all the way from five dollars
> to ten dollars.[27]

Two firms in particular were engaged in the railroad map business, Rand McNally and Poole Brothers. Indeed, in the announcements of the opening of the former's Map Engraving Department in 1872, the attention of railroad officials was particularly sought:

> It is with very great satisfaction that we are able to
> announce the extensive and permanent additions that have
> been made to our engraving department. We are now
> prepared to make sketches and complete the engraving of

31

maps, in the very best possible manner, and at short notice. To those with whom we have had contracts of months standing, to represent their roads and connecting lines by Maps in the *Guide*, we have to say that as soon as the drawings can be approved by the proper officials, the engravings will be completed and the maps inserted in their respective places. The general clearness with which these maps are printed speaks volumes in favor of the efficiency of this department of the house of Rand McNally & Co.[28]

Less than a year later, the editor of the *Railway Guide* reported on the progress:

It is with much pride and a very great deal of satisfaction that we call attention to the constantly increasing number of Map illustrations appearing in the Guide from month to month. The maps of the lines of the Chicago & North-Western, St. Louis & Iron Mountain and Cairo & Fulton railroads in the present number, are referred to as beautiful specimens of the Engravers' Art, and serve well to illustrate the perfection attained in this department by the house of RAND MCNALLY & CO. *To our knowledge, no such relief-plates have ever before been produced in any country;* and while we do not say this work cannot be produced elsewhere, we claim to be in advance of all competitors, and with continued exertion and the facilities at our command, we hope to remain so.[29]

Poole Brothers started as a separate company after George A. Poole had resigned from his post as treasurer at Rand McNally. The firm specialized in the engraving of railroad maps, and their products often appeared in the publications of Rand McNally. As Poole Brothers expanded, they diversified their operation until the map department had become so peripheral that they sold it back to Rand McNally in the late 1950s.[30]

We may identify several reasons why wax engraving was particularly adapted to this characteristic class of maps. In the first place, since each station maintained by the railroad in question had to be shown on the map, with a large number of small names necessary to identify them, the technique's ability to employ small letterpress type was especially significant. The railroad map consisted essentially of lines and points, without the need for halftone. Large printing runs were required, often at short notice. The blocks were often used in a variety of forms to be printed with type, as in timetables in book form or as large wall posters, and consequently had to be produced by a relief process. Finally, the ease with which a large number of corrections could be made on a wax engraving was clearly an asset for this kind of work.

The wax-engraved railroad map took on a character all its own. These maps were usually black-and-white, water-lined, and extremely crowded with names of every railroad station on a particular line, and often the scale and directional relationships were carefully arranged to show the lines and the region served by the company to the best advantage (fig. 2.5). One commentary on this practice of planned distortion was provided in the *Inland Printer.*

SOME RAILWAY MAP-MAKING.

"This won't do," said the General Passenger Agent, in annoyed tones, to the mapmaker. "I want Chicago moved down here half an inch, so as to come on our direct route to New York. Then take Buffalo and put it a little farther from the lake.

"You've got Detroit and New York on different latitudes, and the impression that that is correct won't help our road.

"And, man, take those two lines that compete with us and make 'em twice as crooked as that. Why, you've got one of 'em almost straight.

"Yank Boston over a little to the west and put New York a little to the west, so as to show passengers that our Buffalo division is the shortest route to Boston.

"When you've done all these things I've said, you may print 10,000 copies—but say, how long have you been in the railroad business, anyway?" *New York Herald*[31]

A less journalistic protest came in the form of a letter to the American Geographical Society from the actuary of the Massachusetts Mutual Life Insurance Company:

Dear Sir:
I beg to ask whether it would not be in the province of your Society to remonstrate against a practice in map-making

33

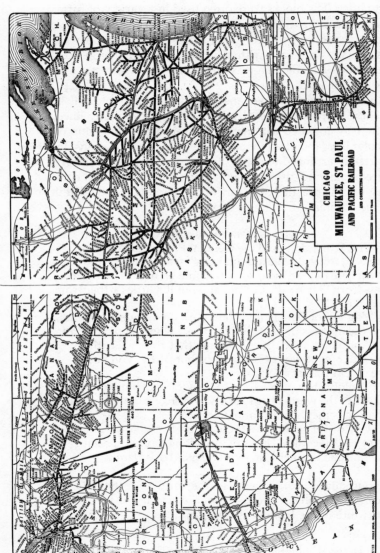

Fig. 2.5. "Chicago, Milwaukee, St. Paul and Pacific Railroad." Compare, for example, the widths of Wyoming and Iowa. From *Atlas of Traffic Maps* (Chicago: LaSalle Extension University, 1930), pp. 134–35

that seems to me quite likely seriously to mislead the young and others in search of information about the geography of our own country. I refer to the practice of publishing maps of the United States in which different scales are used for different parts of the map, without any explanation or obvious line of separation, the two scales in fact being sometimes used in the same part of the same map. The Chicago, Burlington & Quincy Railroad some time ago issued a map, engraved by Rand, McNally & Co., of this character. Taking the distance from Boston to Chicago as a standard, the distance from San Francisco to Denver is shown in reliable maps to be about 1.12 of this standard distance. On the map just referred to it is only about .9 of the standard distance, or 80 per cent of what it should be. The distance from New Orleans to Duluth is shown on correct maps to be 1.25 or about 92 per cent of what it should be. The ratio of the length of the southern side of Colorado to the length of its eastern boundary is really about 1.39; on the C. B. & Q. map it is about 1.21, or about 87 per cent, of what it should be. This map is mounted on rollers and got up in such a manner as to adapt it to use as a wall map, and it is quite good enough in general execution to be preserved for reference.

More recently the New York Central Railroad has issued a map, prepared by the Matthews-Northrup Company, in which the distance from San Francisco to Denver is .64 of the distance from Boston to Chicago, or 57 per cent, of what it should be. The distance from New Orleans to Duluth is 1.04 of the standard distance, or about 76 per cent, of what it should be, and the State of Colorado is almost exactly a square, making the length of southern boundary 72 per cent of what it should be in comparison with the length of the eastern side. This map is issued as a folder, and is finished in very attractive style. It is quite likely to be preserved, and quite good enough in general style to be fastened on a wall for study. Of course cheap sketches of railroad routes are not expected to be accurate as maps, but when our wealthiest corporations issue in large quantities maps of this kind, prepared by the leading map-makers of the country, it seems to me that they are likely to do a good deal in the way of instilling ideas that must be

afterwards unlearned if one is to get at the truth, and that a
protest from the American Geographical Society would be
quite in order, and might be of some effect.

I remain,

Yours respectfully,

Oscar B. Ireland.[32]

There was no response to this elegantly worded tirade, and in
defense of the "distorted" railroad map, it should be pointed out
that the deliberate manipulation of scale to fit a specific function
is an idea well rooted in the history of cartography, going back
at least as far as the Peutinger map, a Roman road map (3d–5th
century A.D.) which distorts direction and scale in order to
squash the entire road network of the Roman Empire (so to
speak) into a map 1 ft. x 21 ft.

The two largest mapmaking firms of the period using wax
engraving were Rand McNally and Matthews-Northrup. Once
again, as in the transitional period, the technique was dis-
seminated through a network of engravers and management
personnel, many of whom had worked at some time for one of
these two companies. Several companies stemmed from Rand
McNally, the firm of Poole Brothers (now the Poole Printing
Company, Chicago), for example.[33] (See fig. 2.4.)

Another connection between Rand McNally and another firm
resulted when George H. Benedict, after serving his apprenticeship as an engraver at Rand McNally, transferred to George F.
Cram as foreman in the production department. Later Benedict
moved again to become manager of the wax-engraving department at Blomgren Brothers Inc., Chicago. All this happened
before 1884, when he set up business for himself at the age of
twenty-seven as the George H. Benedict Company, which continued engraving maps and other subjects until 1903, at which
time Benedict merged with the Globe Engraving and Electrotyping Company, whose emphasis on map work has been
declining ever since.[34]

Around 1880 Caleb S. Hammond joined Rand McNally as a
service representative. He left in 1900 to form his own firm,
C. S. Hammond and Company (now Hammond Inc.), which
has been well known for its maps and atlases.[35]

After the establishment of Poole Brothers, railroad maps soon became a minor part of the business. By the mid-1870s, Rand McNally had begun to produce their famous pocket maps; the first two were the *Guide Maps of Chicago* and the *New Map and Guide to the Black Hills of Dakota*, both published in 1875. In 1876 the company advertised a large wall map of the United States, about five feet by eight feet; this was later published in atlas form as the *Business Atlas of the Great Mississippi Valley and Pacific Slope* (1876–77) and in pocket map form covering some forty different titles by 1878.[36] The *Business Atlas* was to be the forerunner of the annual *Commercial Atlas.*

The company was quick to demonstrate its superiority in the map-printing field. At the World's Columbian Exposition in Chicago in 1893, it exhibited what was said to be the largest map ever printed from a single plate: a 14 1/2 x 9 1/2 feet map of the United States at the scale of one inch to eighteen miles. Near it stood the electrotype from which the map had been printed, all 140 square feet of it. It was apparently made up of pieces so cleverly joined that the joints could not be detected.

Rand McNally continued to use wax engraving until after the Second World War, producing millions of maps and atlases for the mass market, and creating in their name a byword for maps in the minds of generations of Americans. There are few families that do not have a wax-engraved Rand McNally map or atlas somewhere in their belongings, some concealed in history or geography textbooks, encyclopedias, guidebooks, and so on.

Another major map producer during this main period of wax engraving was Matthews-Northrup, a lesser-known name in the map world because much of their work appeared under the imprint of other firms. Yet the company's output was prodigious, and the quality of some of their maps and atlases was not surpassed during this period.

Matthews-Northrup had roots extending back to the 1850s in the firm of Jewett and Chandler. About 1878, William Phelps Northrup, a nephew of Elam R. Jewett (partner of Henry Chandler), after an apprenticeship in his uncle's firm, joined J. N. Matthews to form the company.[37] The firm came to have one of the finest map-engraving departments in the United

States. Two works are particularly worthy of mention. The *Century Atlas,* first published in 1897 by the Century Company under the supervision of Benjamin E. Smith and reappearing in many editions, is a prime example of how wax engraving, in the right hands, could produce clear, crisp maps while maintaining a heavy load of names. Other examples of Matthews-Northrup cartography are the maps in Elroy McKendree Avery's seven-volume *History of the United States,* which one cartographer has called "extraordinary in graphic conception for the time, and . . . a landmark in American map-making."[38] Francis Vinton Greene, in whose book *The Revolutionary War and the Military Policy of the United States* (1911) some of the Avery maps were republished, commented that the maps "have been engraved and printed at the Matthews-Northrup Works in a manner that leaves nothing to be desired."[39]

In 1926 the map department of Matthews-Northrup was acquired by the J. W. Clement Company, and for several years after, maps produced by Clement bore the imprint "J. W. Clement Co. (Matthews-Northrup Division)." In 1963 Clement moved from Buffalo, New York, to a new plant in Depew, New York, and the map department was discontinued. The company had made the transition from wax engraving to offset lithography before 1940, when it adopted the float-lettering method of Colin Landin (a variation of stick-up lettering), the rights to which were sold to a number of companies around 1940, including Rand McNally and A. J. Nystrom.[40]

The Period of Replacement
by Offset Lithography

While offset lithography was commercially introduced by Ira W. Rubel, a New York lithographer, in 1904, wax engraving continued its popularity well beyond that date. This was due to three main factors. In the first place, it took several years for the quality of offset lithography to improve to an acceptable point. Well into the 1930s most printers complained (some still do) of the grayness and "lack of snap" of offset printing compared to letterpress printing such as wax engraving.

Heavy capital investment in plates and machinery was the

second major factor for the persistent use of wax engraving long after it was rendered obsolescent by offset lithography. A vast amount of usable equipment was represented by the electrotyped plates and relief-printing presses geared to the wax-engraving techniques. Map publishers could not be expected to simply abandon this capital and retool for an entirely different printing process. This was particularly true for the printing of maps for which revision was not considered necessary, such as maps in history textbooks. While corrections could be made on the plates, the procedure was not easy; a special class of craftsman arose to handle it. But as long as corrections could be made, the plates were still usable, even for maps that required revising, and this helped perpetuate the technique.

The third factor contributing to the inertia of wax engraving lay in the tradition of specialized engraving skill based on a strong apprenticeship system. We have already observed the closely knit structure of a map publishing industry in nineteenth-century America that was based on wax engraving; the master-apprenticeship system was an important part of this structure. The process was virtually secret, as can be seen from the absence of technical manuals; the only way to learn the trade was to practice it. The apprenticeship took a minimum of two years, and after that period, the apprentice would be allowed to do only the simplest work. In some cases the continued use of wax engraving in a particular firm was largely dependent on the availability of skilled personnel. For example, C. S. Hammond continued to use wax engraving until about 1938, when their plate finisher died, leaving the company no means of correcting or updating their maps.[41] Throughout the 1940s, historical maps continued to be printed in the company's atlases from existing electrotypes, since no new engraving or finishing was involved, while newer maps were printed by offset lithography.

While the technical change from wax engraving to offset could not be effected overnight, the advantages of the latter technique were so evident by the time of the Second World War that few printers and cartographers could argue for the old technique. The versatility of offset lithography was its greatest competitive advantage over wax engraving for the printing of maps. The drawing could be both produced and printed at any reasonable size; and almost any copy, line or halftone, was reproducible by the method. Lithography was cheaper for several reasons, including the frequently underemphasized labor aspects. In the 1930s, instead of paying a trained engraver forty-two dollars to fifty-eight dollars a week,[42] printers could hire girls for about thirty-five dollars to forty dollars a week to make drawings for reduction which were quite satisfactory.[43]

Two major technical improvements related to offset lithography increased its advantages over wax engraving: the introduction of preprinted lettering and the development of scribing. As we have seen, one of the advantages of wax engraving was the ease with which type could be combined with line work. About 1920, the United States Geological Survey began to experiment with a method known as "stick-up" lettering, in which names were printed on gummed paper stock, cut out, and attached to the map manuscript.[44] There had been previous experiments in which type could be transferred to lithographic stones, but the USGS method seems to be the earliest example of stick-up lettering in the sense it is used now. With the development of various photocomposing machines, plastic transparent stripping film, the medium now used today, came into being. These methods provided a convenient way of adding type on maps without resorting to the cumbersome equipment of wax engraving.

The other major development was scribing (the cutting or scraping of an image on a coated surface), first on glass in the 1920s and 30s, then on plastics in the postwar period. Scribing was analogous to wax engraving in the sense that engraving tools were used and a coating was removed, but the training time differed considerably. Draftsmen could be trained to do acceptable scribing work in six weeks, compared to the minimum two-year apprenticeship required for the wax engraver.

The rapidly changing political scene between the world wars and especially after the Second World War exerted pressure on mapmakers to compile completely new atlases. During the 1940s it became obvious to major mapmaking firms such as Rand McNally that the time had come to compile new maps and to prepare them for offset lithography. Yet even into the 1960s, for certain classes of maps, Rand McNally and other companies (George F. Cram, for example) were still making reproduction proofs from wax-engraved electrotypes and then transferring them to offset plates.

No doubt, some pressure to change came also from the competition of those mapmaking firms that had been founded after the introduction of offset lithography, and that were therefore in a position to use the technique from the beginning. Denoyer-Geppert, A. J. Nystrom, General Drafting, and the American Map Company, for example, handled the vast majority of their work by the newer techniques at the outset, since they did not face the problem of having large amounts of capital in the form of relief-printing plates and the machinery to print from them.

On the other hand, at least one major company started just a little too early to avail itself of offset printing. Caleb S. Hammond left Rand McNally in 1900 to form his own company, and wax engraving was the only workable technique known to him for printing maps. Had he started his company ten years later, he probably would have used offset lithography from the beginning. As it turned out, C. S. Hammond used wax engraving until about 1940 for all its maps, and continued to use the technique for historical maps well beyond that date.

The combination of improvement in the quality of offset lithography and scribing, the growing demand for totally new map compilations, and the economies afforded by the new technique led to the disappearance of the skilled wax engraver and electrotype finisher. By 1950 very few firms were using the technique to make new maps, and wax engraving may now be considered an obsolete method for the making of maps.

Wax Engraving in England

Of the many miscellaneous methods of producing a metal relief-printing surface during the 1830s and 40s in England, so well documented by Elizabeth M. Harris, glyphography was to have the most lasting success.[45] Its inventor, Edward Palmer, made no secret of the process. His first patent of 1841, entitled

"Improvements in producing printing surfaces and printing china, pottery ware, music, maps, and portraits," included a section on *electrotint*, a process whereby a molding composition could be applied to a copperplate with brushes and then electrotyped. He made no specific mention of the use of engraving tools or printer's type in the preparation of the mold, either in his patent specification or in a published manual, *Electrotint* (London, 1842), written by Thomas Sampson, Palmer's assistant.

In his second patent of 1842, Palmer described a process very similar to that of Morse's cerography, and which can be considered under the general term of wax engraving. He took a smooth metal plate, either German silver or brass, and after darkening the surface, spread a composition of white wax, spermacetti, resin, and pitch in a thin layer. The wax was then cut through to the plate beneath, built up with extra wax in the areas needing greater depth in the printing surface, and electrotyped. Palmer described the method in another book, *Glyphography, or Engraved Drawing*, which went through two editions in 1843 and a third in 1844. Elizabeth Harris provides a full bibliography of Palmer's glyphography and the books in which glyphographers' prints appeared.

Palmer's glyphography was used for maps in England, but not to the same extent as Morse's cerography in the United States. At least two of Palmer's business associates had a considerable interest in maps. During the early 1840s, Palmer worked out of a small shop in 103 Newgate Street, London, where he sold optical instruments and chemicals. He continued to operate in this shop until 1845. In 1846 he moved to 79 Shoe Lane and opened the "Glyphographic Engraved Drawing Office," with H. G. Collins as manager. Collins had an interest in maps, and his draftsman, R. Jarvis, produced a glyphographed map in a travel book of 1851. Collins eventually gave up glyphography to work on his own process for enlarging or reducing prints by printing them on an elastic surface, stretching them, and offsetting the impression onto paper. The technique, which was known as "electro-block printing," or "Collins' process," formed the basis of a company intended for map printing around 1860.

Another of Palmer's associates, Thomas Best Jervis, set

himself up as a "glyphographic agent" for Palmer in 1845. Jervis had a considerable background in cartography and map printing in India, and although his primary interest appears to have been in lithography, he claimed in a letter offering his services to the governor general of India that he had practiced all artistic means of making maps, and his link with glyphography would imply that he had used that process for maps.

That the process was suitable for maps is not only reflected in the title to Palmer's first patent of 1841, but in the fact that he included a map of the Isle of Wight, engraved by John Dower, in the third edition of his *Glyphography*. By 1855 the *Imperial Gazetteer*, started in 1851 by W. G. Blackie, contained 120 glyphographic maps. The *Gazetteer* reappeared in 1868 and 1873, by which time, while most of the glyphographed maps remained, some had been replaced with maps by Charles Gillot. The firm of Chapman and Hall also employed glyphography for many of their maps, including a special series sold at a penny apiece.

In 1872 a patent was granted to Alfred and Henry Thomas Dawson for a process that was essentially the same as Palmer's. At this time, several refinements had been added to the art. The plating dynamo (powered by steam) had been invented by Woolrich (1842), replacing the earlier Smee battery (1840) for which thirty or forty hours of electrodeposition were necessary. The plating dynamo could deposit a sufficiently thick shell in two hours. In addition, negative photography could now be used to transfer the image to the wax plate. The Typographic Etching Company of the Dawson brothers produced wax-engraved plates on a successful commercial scale until switching to process engraving, specializing in color work, at the end of the century.[46]

While referring to the Typographic Etching Company, it is appropriate to recognize the cartographic work of Emery Walker, who was to become a well-known engraver and printer in Britain, particularly through his association with William Morris. Walker had joined Alfred Dawson's Typographic Etching Company in 1873 as a pressman, and he undoubtedly learned wax engraving there. He was to use the technique later

Chapter Two

for the monochromatic maps in the 11th edition of *Encyclo-paedia Britannica*, the maps in L. Dudley Stamp's *The World: A General Geography* in its many editions, and no doubt for many other maps in books.[47]

Maps formed a very small part of the total output of wax engraving in Britain and indeed in Western Europe. The process, whether under the name of glyphography, electrotint, galvano-glyphy, typographic etching, and a myriad other names, had been used mainly for line drawings. It was, after all, designed to allow freedom to the illustrator. For maps, the tradition of lithography and copper engraving was extremely strong, and the use of wax engraving for large commercial atlases never became a reality in Europe.

WAX ENGRAVING IN EUROPE AND
AMERICA COMPARED

When the experience of the United States and Europe with wax engraving is compared, the question arises why the technique was almost entirely confined to the United States, while the traditionally renowned centers of cartography in Europe hardly used it except for small maps in books, periodicals, and news-papers. The answer seems to lie in the fact that European cartography became industrialized later than American owing to three closely interrelated factors—technical, aesthetic, and economic.

There had been a 300-year tradition of copper engraving in Europe that even lithography took a century to break. Many European mapmaking concerns were still engraving maps on copper at the beginning of the twentieth century; some highly specialized branches of the industry still do. There was therefore a considerable investment of equipment and skilled labor that the Europeans could hardly be expected to abandon in favor of a process requiring different engraving techniques, tools, plate-making equipment, and printing machinery. In addition, the European cartographer had been raised in a powerful tradition that exalted both great traditional skill and an aesthetic sense. While the nineteenth century was a time when draftsmen practiced their skill by writing the Lord's Prayer on something the size of a postage stamp, it was also a time when they were

44

expected to have an artistic flair in such tasks as, for example, the positioning of lettering. To the European traditionalists, *all* lettering was to be done by hand; type simply had no place on a map. This tradition was in large measure due to the powerful role of the geographical societies, such as the Royal Geographical Society and the Société de Géographie de Paris, in setting standards, and the family pride in the European "geographical establishments" such as Bartholomew's, Stanford's, Perthes, and others which held quality at a premium. Much of the emphasis on quality was also due to strong traditions in the established government mapping agencies: the Ordnance Survey, the Reichsamt für Landesaufnahme, the Institut Géographique National, and many others.

One imagines that, when type finally appeared on American wax-engraved maps in any significant amount in the 1870s, it was greeted by these European agencies as another example of crass American commercialism. The type certainly was far from aesthetically pleasing, and its avoidance by Sidney E. Morse during the early days of wax engraving in the 1840s may have been one attempt on his part to emulate the delicacy of copper engraving.

Criticisms of American commercial atlases in the literature started at least as early as 1915. In that year, there appeared a letter in the *Inland Printer*, from which extracts are reproduced here:

> I readily admit that the maps of the United States Geological Survey are very fine and accurate maps, but many of those issued by the various publishers lack all of these qualities. They are not at all pleasing in appearance, the state maps, etc., resembling very badly colored checkerboards, usually out of register somewhere, and very inaccurate. This you can easily control by comparison with the United States Geological maps. The same is true for the school atlases. You may think very much of typesetting for river and town names, etc., but it has many disadvantages when compared to the free-hand lettering, and is not nearly as beautiful.
>
> . . . .
>
> I sincerely hope that our mapmakers take the footsteps of those foreign mapmakers, for the type-set maps of the

process described in your June number certainly are no match for those lithographed or copper and steel etched maps.[48]

Toward the end of the 1920s, the differences between European and American atlas maps were highlighted by the debate over certain tariff loopholes being exploited by at least one American atlas publisher, who had the maps for his publication printed in England and sent over to the United States, where they were bound, copyrighted, and marketed. The president of the American atlas publisher involved, John W. Hiltman of D. Appleton and Company, communicated the following view to *Publishers Weekly:*

> We still believe, along with many experts in this country, that American map producers cannot attain the high standard of maps required for school use at a price to make the product available for educational institutions. To hamper the use in schools of geographical tools of the highest quality by means of a tariff barrier will penalize American education.[49]

Also leveling his criticism against American map producers was Eugene Van Cleef of the Department of Geography, Ohio State University:

> For years, American geographers and educators generally have recognized the superiority of European map-makers over the American . . . we have never produced an atlas in America equal to the best produced in Europe. When we want the last word in an atlas, we seek either an English, German, or French product.[50]

Such English publications as might have reviewed contemporary American atlases, for example, the *Geographical Journal*, indicated their disdain for such products by totally ignoring them in volumes 1–68, until their review of *Goode's School Atlas* in 1927, which concluded with the following remark:

> Notwithstanding some defects the Atlas is a notable piece of work if only for the fact that it is, as far as we are aware, the best which has yet been produced in the United States.[51]

As late as 1950, an unnamed reviewer of eight atlases in the *Geographical Review* made explicit a sentiment that had probably been in the minds of many:

> Commercial atlases in the United States have long been inferior to the better atlases produced in Europe in design, selection of color, printing, and, not infrequently, accuracy. Perhaps this is due to inertia or lack of map education on the part of the American public, which has failed to create a demand sufficiently strong to justify the heavy expenses of an entirely new compilation. Whatever the reasons, this reviewer cannot believe that the general run of American atlases has reflected the best efforts of geographers, cartographers, and printers in the United States.[52]

From these quotations, it is clear that, not only was there a characteristic difference between the style of American and European maps, but that the latter were aesthetically far superior in the minds of informed consumers. There is therefore little wonder that European cartographers did not use wax engraving more than they did: the products of the technique were demonstrably inferior to their own.

There were economic reasons why wax engraving did not assume more importance in Europe. While more research needs to be done on this matter, it seems that European press runs for maps were considerably less than their American counterparts, even in commercial cartography. Thus, the speed of the rotary offset press was not immediately impressive to European map producers, whereas according to Koeman, the offset press fulfilled a long-entertained desire in America to increase production speed.[53] If speed had not been achieved in the American use of wax engraving, the length of printing run certainly had. Plate life was far longer than in copper engraving or lithographic printing from stone, and the plate could stand up to the rigors of the power cylinder press.

The lower wages of the European cartographer helped to offset the greater costs of production demanded by copper engraving or lithography. By the 1920s, there was a tariff of twenty-five percent on imported European maps for fear they would flood the American market. In the Senate hearings on these tariffs, one leading American map publisher requested that a 75 percent tariff be placed on all such atlases *in addition* to the 25 percent already in effect, on the grounds that the wages paid in the United States were more that twice those paid in England for the same work.[54]

What emerges from this discussion is that the character of the American map publishing industry was far more industrialized or organized on mass production lines than its European counterpart; that the quality of the product suffered in this mass production atmosphere; and that the European map publishing houses placed more emphasis on the artistic aspects of cartography, which they considered could be attained more easily by the traditional processes.

A number of possible reasons may be proposed why the atlases of certain American atlas publishers such as Samuel Augustus Mitchell (1792–1868) and George W. Colton (1827–1901) were copper or steel engraved and not wax engraved. S. A. Mitchell continued to publish from the Henry S. Tanner establishment which he had taken over in the mid–nineteenth century. H. S. Tanner's main work had been produced before the invention of wax engraving, and thus was engraved on copper; Mitchell continued to use the available plates and equipment.[55] In addition, the firms of Mitchell and Colton had chosen to widen the market for their products, not by tapping the cheap atlas market in the United States, but by entering the European atlas trade. It was thus essential for them to match the fine quality of the intaglio atlases of France, Britain, and Germany.

The use of wax engraving did not extend to county atlases, fire-insurance maps, and automobile road maps for several reasons. In the case of county atlases and Sanborn fire-insurance maps, the editions were usually very small, not warranting the use of a relief-printing plate on a power press. While the production of county atlases was an extremely important chapter in the history of American cartography, yielding well over four thousand different atlases and plat books, an edition of a thousand was considered better than average.[56] The total circulation, consequently, probably did not exceed four million.

Similar to the type of market served by the county atlases was that of the Sanborn fire-insurance maps, produced in very small editions, and notwithstanding the large number of separate publications of considerable value to both contemporary businessmen and to present-day researchers, did not reach the general public in large numbers.[57]

The road map was an anomaly for a different reason. This type of map, produced first for bicyclists and then for automobile owners, had not become a popular item until the first decade of the twentieth century. By the time of the First World War, when the road map business had become too big for all but the very largest mapmaking concerns, the quality of offset lithography had also become sufficiently improved to attract this work. Since the road maps were essentially new compilations, the "big three" (Rand McNally, General Drafting, and H. M. Gousha) were able to start afresh with offset lithography.[58]

3 The Techniques

The foundation of the wax-engraving process was a flat, smooth surface on which wax could be flowed for engraving. This foundation was called a *case* or *plate* and any imperfection in it would be reflected in flaws in the finished printing plate.[1] Almost any flat material could be used, provided it could be polished to a smooth surface, and provided it was inexpensive and easily available. Glass, steel, brass, silver, wood, and albata[2] are all mentioned in the literature, but by far the most common material for the case was hard engraver's copper.[3] Copper was strong, easily polished, and usually readily available during the peak years of wax engraving. Since the cases were used repeatedly, the initial investment could be amortised over a long period of time, and the initial cost was not particularly high. In the 1930s, a 9 x 12 inch copper case cost about five dollars.[4] The limit to the size of the case was dictated by the size of the electroplating bath in which it was to be immersed. In thickness, the plate varied between one twenty-fifth and one-sixth of an inch, depending on the size of the engraving.[5] A larger engraving required a thicker case to maintain rigidity.

Cases were thoroughly cleaned before each use. The remnants of old wax were scraped off with a piece of cardboard while the plate was heated over a gas flame. Then the engraver used a rag to clean off much of the rest of it. The case was then washed and scrubbed with ground pumice and a wet cloth or piece of thick felt to remove further remnants of wax or the previous coating of silver nitrate. If it required further cleaning, it was washed with a solution of lye or potassium cyanide, a highly poisonous substance that required great care in handling. One engraver relates how he and his fellow apprentices would pretend to eat the cakes of the compound, to the horror of their superiors.[6]

The clean copper case was stained dark, usually with a solution of silver nitrate or sulphate of potassium or copper.[7] The purpose of coating was to prevent the electrodeposited copper from bonding to the copper case during electrotyping, and to provide a contrast with the white or yellow wax when the mold was engraved. This contrast not only allowed the engraver to see clearly what he had done, but also enabled photostat proofs to be taken of the engraved mold before electrotyping, so that these could be read against copy and last minute changes made in the wax.[8]

THE WAX

Engraving wax consisted of white beeswax mixed with a variety of ingredients to satisfy the following requirements: good adhesion to the case; a clean cutting quality, yet not so hard as to be fragile; ability to retain its consistency for a long time without drying out and cracking; a homogeneity or smoothness to avoid minute inconsistencies which could affect the evenness of the line work and lettering.[9]

In the early years of wax engraving, the formulas for the wax molding compound were jealously guarded secrets. Each engraver had his own, and he was sure that his was the best.[10] Consequently, there was much variation in both the ingredients and the proportions in which they were mixed. In addition to a variation from engraver to engraver, different formulas were used for summer and winter conditions. Table 1 summarizes the variations. The three basic ingredients for the molding compound were beeswax, Burgundy pitch (a resinous product obtained from the bark of the Norway spruce tree), and zinc oxide. The pitch (sometimes mixed with resin) acted as a binding agent, while the zinc oxide, as well as giving a white and opaque property, also served to harden the compound. Other ingredients included turpentine (especially Venice turpentine, a fine clear substance, obtained from the larch tree), which was used for thinning the compound when necesssary.

TABLE 1. ENGRAVING WAX FORMULAS

Source	Total	Wax	Pitch
Reinhold, 1886	5 1/4 oz.	Beeswax 76% (4 oz.)	Venetian pitch 19% (1/4 oz.)
Inland Printer, 1895a		Beeswax 12.5% White wax 17%	Black pitch 1% Burgundy pitch
Benedict, 1912		Beeswax (bleached)	Burgundy pitch 29%
Sherman, 1915		Beeswax 75%	Venetian pitch 20%
Rand McNally, 1965 (summer formula)	77 oz.	White wax 48 oz. Paraffin 1 oz.	Burgundy pitch 5 oz.
Rand McNally, 1965 (winter formula)	73 oz.	Paraffin 1 oz. White wax 48 oz.	Burgundy pitch 3 oz.
Schwartz, 1930		White beeswax	Burgundy pitch
Dawson, 1872		Paraffin 75%	
Servoss, 1906		Paraffin	
Palmer, 1842	5 oz.	White wax 40% (2 oz.)	Burgundy pitch 20% (1 oz.)
Mitchell, 1968		Pure white beeswax Candilla wax	Burgundy pitch White pine pitch

a "Preparation of Wax for Wax Engraving."

The waxes and resins were melted separately and then combined in a single pot. Zinc oxide was then added to the compound. It was necessary to grind the mixture with a bone muddler for eight hours in order to remove the inconsistencies caused by the fine crystals of zinc oxide. Originally this grinding was done by hand in pharmaceutical fashion, but machines eventually took over this part of the process. The molten wax was poured into molds about 1 x 4 x 8 inches and left to cool. The resulting slabs could then be stored prior to use by the engraver.[11]

Resin	Spermacetti	Whitener (Zinc, etc.)	Other
		Oxide of zinc 5% (1 oz.)	
1%	34%	White zinc 34%	
Pale Resin		French dry zinc	
		Oxide of zinc 5%	
6 oz.	1 oz.	Zinc oxide 16 oz.	
4 oz.	1 oz.	Zinc oxide 16 oz.	
Rosin		Zinc Oxide	Venice turpentine
		Tungstate of lead 25%	
		Zinc white	
Rosin 20% (1 oz.)	20%	Sulphate of lead (dusting)	
		Zinc oxide	Venice turpentine

LAYING THE GROUND

In order to obtain a thin, even coating of wax on the copper case, the latter was placed on a hotplate and a slab of wax was rubbed over it, as demonstrated in figure 3.1. Occasionally the wax was dumped or poured on the center of the plate and rocked to spread it evenly to the edges. The spreading was often assisted with a comb or rakelike tool.[12] After coating, the plate was chilled on a level stone slab to harden the wax.

In order to ensure a correct, even thickness of the wax ground, the wax was scraped by the engraver with a flat piece of steel.

Fig. 3.1.   Laying the wax ground. J. Stanley Haas at the Haas Wax-Engraving Co.

This was a critical part of the process and was not a job for inexperienced apprentices.[13] The desired thickness of the wax varied according to the coarseness of the engraving. Relatively coarse subjects such as ruled forms were classed as "deep" wax engravings, while other types of finer diagram and map work were known as "thin" wax engravings. The advantage of deep wax engraving was that less building was required between the engraved lines to give depth to the electrotype. Generally speaking, the deep engravings had a larger proportion of type matter in relation to line work than the thin engravings.[14] Table 2 gives the ranges of thicknesses as suggested by several authors. The thickness of the wax ground could be measured easily with a *halftonometer*, or Levy halftone gauge, originally designed to measure the depth of halftone-printing plates to within .0005 inch (fig. 3.2).

The Techniques

TABLE 2. Recommended Thicknesses of Wax Ground

Source	Thickness (inches)
American Dictionary of Printing and Bookmaking (1894)	1/8 (extremely coarse)
Bormay, 1900	1/20
Sherman, 1915	1/24
Mitchell, 1965	1/32
Inland Printer, 1902[a]	1/50 (for maps)
Landes, ca. 1920	1/64
Dawson and Dawson, 1872	1/80
Benedict, 1912	1/100 (coarse)
Rand McNally, 1965	1/125 (coarse)
Benedict, 1912	1/200 (fine)
Rand McNally, 1965	1/250 (fine)

[a]"Cereography or Wax Engraving."

TRANSFERRING THE IMAGE
TO THE WAX

In all but the simplest wax engravings, it was necessary to transfer the image of the subject to the surface of the wax as a guide for the engraver, and to enable him to check the progress of his work. The methods ranged in sophistication from simple hand sketching in the wax to the photographic reduction of a larger drawing onto the sensitized wax surface. In hand sketching, the image was sketched lightly in the wax, care being taken not to engrave through to the case. This straightforward method was used in the early days of wax engraving before the introduction of more sophisticated techniques, but it was occasionally used for simple diagrams and ruled-form work throughout the history of the technique.[15] Where reduction or enlargement was desirable, a pantograph was employed, or square copying grids of different sizes were drawn on the wax and the copy, and a proportional drawing made on the wax.[16]

Another method used the well-known carbon paper principle: a piece of carbon paper was laid between the copy and the wax surface, and a tracing made. In a variation of this method, the back of the draftsman's copy was covered with lithographer's red chalk, and the copy placed face up on top of the wax and

55

Fig. 3.2. Levy halftonometer. Courtesy Rand McNally and Co. Photo, Peter Weil

fixed by holding a warm iron at each corner so that the wax melted and adhered to the paper. Then the tracing was made by going over the detail on the copy with a steel point and leaving a red chalk impression on the wax.[17] In addition, an inked impression from a zinc-etched line block or other printing plate could be taken on the rubber blanket of an offset press, printed on paper, and transferred by pressing it while wet to the wax surface.

Photography improved these methods considerably, rendering the transfer process both quicker and more accurate. The normal method was to reduce the copy, base map, or compilation to engraving scale photographically to produce a negative. The surface of the wax was sensitized with a solution of albumen, zinc oxide (or zinc carbonate), and silver nitrate and allowed to dry in a darkroom. The negative was then placed on the wax surface and exposed to light, printing a positive, right-reading image. After developing, fixing, rinsing, and drying, the wax plate was ready for the engraver.

ENGRAVING

The task of the engraver was to produce an intaglio mold in the wax ground from which a printing plate could be cast. It was essential that all points, lines, and symbols be cut cleanly through the wax to the oxide stain protecting the case, so that the printing surface of the electrotyped plate would be suitably level. Care was required to avoid disturbing the oxide stain, as the electrodeposited copper would then bond with the copper case, and the shell would be destroyed in an effort to separate them. The technique of stamping type and symbol dies into the wax was quite distinct from engraving, and the two phases were usually executed by two different groups of craftsmen.[18]

Although some of the tools used by wax engravers were made and sold by commercial manufacturers, the large majority were made by the engraver himself. Indeed, wax engravers often took almost as much pride in fashioning their own tools as in the finished engraving, a feature which lent much of a "craft-industry" or homemade air to the commercial process of wax engraving.[19] Many of the hand-engraving tools were made from dentists' picks and drills, sewing machine needles, or darning needles, inserted in wooden handles or penholders and held rigid with sealing wax.

Point Symbols

Cartographic point symbols, such as those for settlements, were engraved on soft wax with commercially produced dies. The wax plate was placed on a warming block until pliable. The dies were moistened to avoid their adhering to the wax, and pressed by hand through the wax to the case beneath. It was clearly essential that the tool be perpendicular to the plate while stamping, thus ensuring that every part of the symbol would print evenly when converted into the electrotype.[20]

Chapter Three

Lines

The basic tool for engraving irregular lines by hand was the *graver*, similar in appearance to the wood- or copper-engraver's burin. It was a V-shaped tool with a groove ground down the center to act as a diverting channel for the wax burr (shaving) resulting from the engraving. Typical graver points are illustrated in figure 3.3. Unlike engraving in wood or copper, very little pressure was required to cut lines in wax, with the result that the engraving point for a wax-engraver's tool was set in a long thin handle, a contrast to the copper- or wood-engraver's burin, the handle of which was shaped to fit into the palm of the hand to give extra support for the application of pressure. About ten or eleven gauges of tools were normally used by the engraver, to give a range of line thicknesses from a hairline to about 6 points (approximately 6/72 inch). The V-shape for wax-engraving tools was by no means the only one used. The tools could be lozenge-shaped, flat, or pointed, and Dawson describes a tool with a triangular section similar to that on the far left in figure 3.3. Double-line tools, similar to the modern parallel-line scriber, were used to engrave parallel roads or railroad casings. These tools were made ingeniously by cutting across the eye of a needle and grinding down the two prongs to the desired shape. Boundary lines made up of dots, dashes, or other symbols were engraved with wheel tools which had the appropriate symbols on the periphery.[21]

The basic V-shaped engraving tools were pushed away from the engraver so that the excised wax slid up the groove in the graver and out of the engraved line. The resulting shavings or "burrs" were dusted off the engraving from time to time with a soft brush.[22] Rivers were often cut with a thin V-shaped tool at the source, graduating to thicker tools downstream, a procedure which has its counterpart in modern scribing.[23] Hachures were often engraved in a similar way, although Rand McNally sometimes used stamping tools with various standard arrangements of hachures.[24] The wheellike tools, used for engraving boundary lines consisting of an assortment of symbols, were rolled through the wax in "map-measurer" fashion, care being taken to keep the symbols on the edge of the wheel cutting right

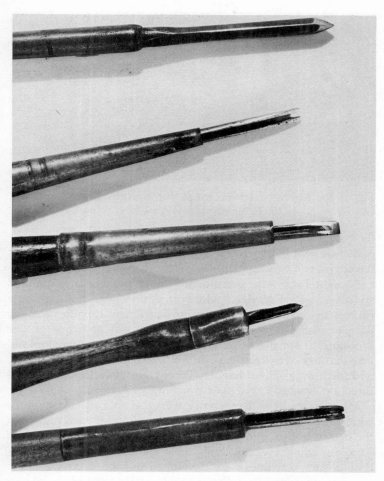

Fig. 3.3. Graver points. Courtesy Rand McNally and Co. Photo, Peter Weil

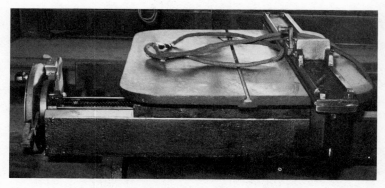

Fig. 3.4. Ruling machine made for Buffalo Wax Engravers, Inc., around 1920. Courtesy Haas Wax-Engraving Co.

through the wax to the case beneath. Throughout engraving, the plate was kept lightly warm and was often tilted at an angle on the engraver's bench.

Straight lines were occasionally hand ruled with a pointed stylus and a straightedge.[25] More often, they were engraved by a ruling machine, a simple yet precise instrument capable of ruling parallel straight lines in any direction to an accuracy of one thousandth of an inch (figs. 3.4 and 3.5).[26] The engraving tool slid on a carriage over a flat base or bed which could be moved back and forth by a geared handle. To engrave parallel lines in any direction, the bed or the plate was simply swiveled.

More sophisticated machines were capable of engraving a wide range of mathematical curves by simultaneously moving the bed in both x and y directions with two separately geared handles, while the engraving tool remained stationary. Although such machines were developed largely for engravers in wood, copper, and steel, they could be easily adapted to the wax engraver's frequent tasks of complicated dial charts and mathematical diagrams.[27] The methods of engraving both straight and curved lines by machine were essentially the same as those for hand engraving, except that the tool was set in a rigid toolholder fixed to the machine in such a way that it would cut exactly to the right depth without disturbing the oxide stain. (Fig. 3.5)

Fig. 3.5. Ruling straight lines on a wax plate. Courtesy Haas Wax-Engraving Co.

Tones

Fine tints were ruled with a ruling machine as described above. Theoretically, such a machine was capable of ruling lines one five-hundredth of an inch apart (.002 inch),[28] but the problems of keeping the wax intact between such a fine ruling outweighed the advantages. In addition, the difficulty of making an electrotype containing such fine lines, and printing from such an electrotype, precluded the ruling of tints denser than 150 lines per inch, the usual number being 100–150 lines per inch.[29] This was quite fine enough for general purposes.

In ruling line tints, the plate was held on the engraving bed by two clamps moving in slots (fig. 3.5), and the bed of the machine was kept heated from below to maintain the pliable quality of the wax. The tool was moved back and forth across the wax plate, while the bed was moved one increment each time by turning the preset gears the required amount.[30]

Ruling flat tints was a satisfactory way of screening down solid color, and a large number of color combinations could be obtained by superimposing tint rulings of primary colors, combining, for example, yellow and blue to produce green.[31] Solid tone was produced by engraving around the area desired, chilling the plate on a block of ice, and chipping away the wax from the engraved boundary line. The rest of the wax in the area could then be scraped out.

The usual method of producing flat stippled ("dotted") tones was to use a graining tool consisting of a wheel fitted with several spikes. This tool was rolled over the surface of the wax either by hand or in an engraving machine to produce a series of parallel dotted lines. Graduated stippling, by varying the size or density of the dots, was done by hand with a pointed stylus, each dot requiring separate placement. It should be pointed out that photomechanical halftones were not possible with wax engraving. When they were desired, it was necessary to make separate plates by photoengraving and to print these in conjunction with wax-engraved plates at the printing stage.[32]

Lettering

In the early days of wax engraving, hand engraving of lettering was common, as in the maps of Sidney Edwards Morse. As the process developed and type became widely used, hand lettering became uncommon in the technique, except for flamboyant letterheads, title lettering, and the adding of embellishments to lettering stamped in with type.[33]

The use of type was the most important competitive advantage of wax engraving in the preparation of maps, diagrams, and other classes of work which consisted of a combination of line work and lettering.[34] The stamping of type through the wax was probably not quicker than hand lettering except in certain circumstances. By the time the type had been set up by hand, arranged in a stamping stick, positioned, and pressed in, a

TYPE INDEX — Italic Numbers — CAPS only

	2½	3	4	5	6	7	8	9	10	12	14	16	18	20	24	30	36	48
Alternate Gothic 3					6		6		6	6	6		6					
Alternate Gothic 2					6		6		6									
Alt. Goth. Italic										2	2		2		2			
Announcement Italic							6		6	6	6		6		6			
Bernhard Goth. Italic Med.							2		2	2	2		2					
Bernhard Goth. It. Light					2		2		2									
Bodoni													1		1	1	1	
Bodoni Open															8		8	
Bodoni Italic									8	8	8		8		8			
Bookman Old Style													4	4	4	4	4	4
Caslon Bold															6	6	6	
Caslon Bold Italic											1		1		1			
Caslon Light Cond.										4	4		4		4		4	
Century Bold					5		5		5	5	5		5		5	5	5	
— — Italic										1			7			1	1	1
— Schoolbook					5		5		5	5	5		5					
— — Italic					4		4		4	4	4	4	4		4		4	
— Expanded					6		2											
— — Italic	1																	
Cheltenham Bold Cond.											3		3		3	3	3	3
Clarendon Light			4															
Clearface Gothic						3			3	3	3		3					
Cloister Italic Swash												7						
Copperplate Heavy													8					
— Light										8								
— Cond.										8								
Garamond - American							3		3	3	3	3	3					
— Bold							7		7	7	7	7	7					
— Italic Swash									7		7							
Gothic 544				7	7		7		7	7	7	7	7					
— 545	4	7		7	7	7	7	7	7	7	7		7		7	7		
— Mono							7		7									
Kaufman Bold (Italic)													5		5			
Light Line Gothic				7	7		7		7	7			7					
Lining Antique					5		5		5		5		5		5	7		
Lydian									6	6	6		6					
— Bold									3	3	3		3			3		
— Italic									6	6	6		6		6			
News Gothic							3		3	3	3		3		3	3		
— Condensed	2		2		2		2		2	2	2		2		2			
— Extra Cond.					4				3									
— Bold											3							
— Italic				2	2		2		2	2								
Piranesi Bold (Italic)							5		5	5	5		5		5		5	
Roman 20				4														
Roman 590 Italic					4													
Spartan Black									1	1								
— Heavy																1		
— Medium			1		1		1		1	1			1		1			
— Med. Italic			1		1		1		1	1			1					
Stymie Bold				8	8		8		8	8	8	8	8			8	8	
— Light					8		8											
— Medium				8	8		8	8	8	8	8		8					
— Med. Italic				4	4		4		4	4	4	4	4					
20th Cent. Med. Cond.							6		5									
Webb Type										8	8		8			8	Cut	8

Fig. 3.6. Sizes and styles of type available at the C.S. Hammond Co. in the 1930s and 1940s. The numbers in the table refer to the cases in which the type could be found. Courtesy Ferdinand von Schwedler

Fig. 3.7 Curved stamping stick. Courtesy Ferdinand von Schwedler. Photo, Peter Weil

skilled hand engraver generally would have finished the job. But for smaller sizes of lettering, in the 4- to 8-point range, type was almost certainly quicker, especially after the introduction of mechanical composition.[35]

Stamping may not have been quicker than hand lettering, but it was easier to do, and easier to find people to do it. While stampers were extremely skilled technicians through long apprenticeship and experience, men of this bent were more available than lettering artists.

Stamping Tools

Wherever possible, clean, new, unblemished type was used for wax engraving, so that the end result would be as sharp as possible. Foundry type was preferred over Linotype, because it was made of a harder alloy that would not be damaged under pressure, but machine-set Monotype was sometimes used. For work that demanded type to be pressed in extremely close to line work and other lettering, the shoulders of the foundry type were shaved down to avoid damage to the adjacent engraving.[36]

For the average-sized map, Sherman estimates that one to five

Fig. 3.8. Straight stamping stick. The feet in the corners are spring loaded and aid precise vertical motion. Courtesy Rand McNally and Co. Photo, Peter Weil

galleys of type were required.[37] Of this, a large amount was in very small point sizes, such as 4- to 6-point, and a great variety of styles were available. This is illustrated by the type index used at C. S. Hammond in the 1930s and 40s (fig. 3.6). The numbers in the body of this table refer to the typecase in which each style and point size could be found. Note that Gothic 545 was available in 2 1/2-point capitals.

The engraver responsible for the lettering (or stamper, as he was called) nearly always set his own type and kept it in a common printer's typecase. The composed names for each job were laid in small sorting racks, so that each word could be easily picked up and locked in the stamping stick.

The stamping stick could be curved, as in figure 3.7, or, more usually, straight (fig. 3.8). Some were two-handled to afford better control,[38] and some were combined with a frame which rested on the wax while the name was being stamped, to ensure verticality. Machines were designed to stamp lettering. The more usual form of stamping machine was somewhat similar to the ruling machine described, except for the important difference that the engraving tool was replaced by a type holder. The type holder was designed to swivel into an upright position to receive

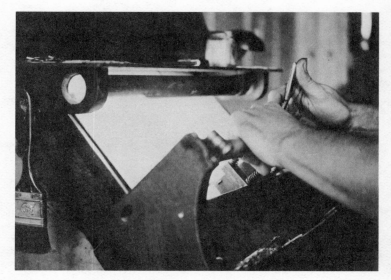

Fig. 3.9.   Machine stamping. Courtesy Wallace B. Mitchell

the words of type, which were locked into it. The holder could
then be swiveled 180 degrees and the type pressed into the wax
by a spring lever (fig. 3.9). Another machine, designed and built
by J. Stanley Haas, had the advantage that the type could be
pressed down with a foot pedal which moved the entire carriage
and type holder in a vertical plane. In addition, the depth to which
the type could be impressed was precisely controlled by checks on
the machine.

Stamping Methods

In map work, the stamping of names was usually done after the
line work had been engraved. An exception to this was the
practice of Rand McNally, where stamping was the first step in
engraving the plate. The type for the job in hand was set up in
composing sticks and transferred to galleys, and proofs were
pulled and checked before the type was given to the stamper. It
was necessary to warm the plate to 85°–110° F., often over a
small tank of heated water, in order to soften the wax. Hard wax
could easily damage type, or at least cause undue wear. To
avoid having the type sticking to the warm wax, the type was

moistened. The tongue was best for this purpose, but those engravers who were unwilling to lick the type for various reasons used soapy water from a small sponge in a dish close by. Mitchell's view is that soapy water was strictly second best.

If words were widely separated, or if the curvature of a name was such that it could not be accommodated in a stamping stick, each type slug was "fingered-in" separately by hand, between the thumb and forefinger, great care being taken to keep it vertical.[39]

In hand stamping, the type was locked in the holder, moistened, pressed down vertically, and rocked slightly to ensure that the type was perfectly flush with the case (fig. 3.10). Stampers worked from the top of the engraving to the bottom to avoid damaging previous work with the hands or tools.[40]

Machine stamping was essentially the same, except that the holder was automatically squared to the wax surface. The holder was swiveled into an upright position, and the type locked in the holder; the holder was then swiveled to a downward position, positioned accurately for the name, and pushed down through the wax. Occasionally, stamped lettering was embellished by hand to give it a less mechanical look.[41] The long-serifed lettering found on many wax-engraved maps published by Rand McNally bear witness to this procedure.

Corrections in Engraving

At this stage, when the engraving was essentially complete, the wax plate was carefully checked for errors with a magnifying glass. Should any be found, a *mender* (a simple, blunt, heated copper instrument) was used to melt the wax in the offending area and smooth it to its original thickness. On hardening, the wax was reengraved with the correct version. If a large area required mending, small pieces of wax were laid on the plate and melted over the errors.[42]

Building

It is clear that the engraved lines and stamped letters in a wax coating of about one-fiftieth of an inch thick or less would produce a printing plate of insufficient depth for letterpress printing. Were such a mold electrotyped and used for printing,

Fig. 3.10.   Hand stamping. From *Inland Printer* 55 (1915): 323

TABLE 3. BUILDING WAX FORMULAS

Source	Amount	Wax	Thinners	Binder	Zinc Oxide
Palmer, 1842		White wax	Turpentine		
Rand, McNally, 1965	78 oz.	Beeswax 23 oz. Paraffin 46 oz.			Zinc Oxide 9 oz.
Reinhold, 1886	11 oz.	Yellow wax 73% (8 oz.) Paraffin 18% (2 oz.)		Venetian pitch 1 oz. 9 oz.	
Benedict, 1912		"Thickened beeswax"			
Inland Printer, 1895a		Beeswax 30% Parrafin 20%		Resin 25% Asphaltum 25%	

a "Preparation of Wax for Wax Engraving."

the spaces between the raised portions of the plate would also catch ink and print. These areas had to be deepened in some way. The standard procedure was to *build up* the spaces between the engraved lines with extra wax on the mold, thus forming hollows in the electrotype cast from it. These hollows were known as the *hold. Building* or *building up* was also known as *bridging* or *making a hold.*[43]

Building wax or *fill-in wax*[44] usually consisted of the waste wax retrieved from old plates, mixed with yellow beeswax. Old wax was usually too stiff, and the right amount of beeswax to be added to give it the right consistency could only be estimated by experience.[45] Building wax was generally of a heavier consistency than engraving wax, and it was often tinted to contrast with it.[46] Different formulas are given in table 3. The wax was made in sticks, about a quarter of an inch square and six or

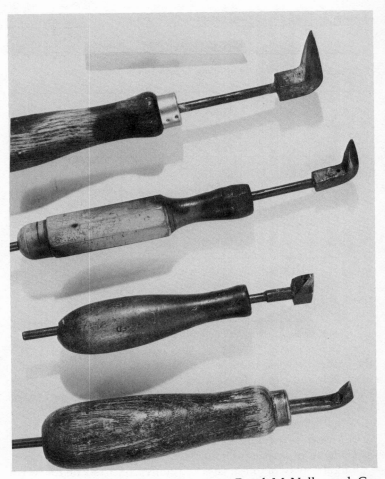

Fig. 3.11. Building irons. Courtesy Rand McNally and Co. Photo, Peter Weil

seven inches long. These could be used conveniently with a building iron (explained below). When brushes were used, the building wax was melted and poured into small pots.

The basic tool for building was the *building iron*, or *fountain tool*, shaped like a small soldering iron, with the tip kept hot

Fig. 3.12.   Building a wax plate at the Back Bay Electrotyping Co., Boston, 1937. Courtesy Wallace B. Mitchell

throughout the operation by a minute flame fed with gas. An earlier prototype of the building iron was the *building pen*. This consisted of a tight coil of copper wire with two copper wires wedged in the coil to form a point. The wax was held in the coil by capillary action, and its flow could be started by bringing the tool in contact with the wax plate. Another variation of this tool used a copper cylinder, a quarter of an inch in diameter, with a groove cut in one side, and wound with copper wire. The wire was connected to an electric current for heating the tool.[47]

The building iron was heated initially in a gas burner and kept hot by the small flame at its head. A stick of building wax was kept in contact with the iron over the wax plate, and the melted wax was allowed to run onto the engraving. All spaces more than "a few hundredths of an inch wide"[48] were built up in this way; when the spaces were narrower, it was impossible to keep the building wax from fouling the lines, and these spaces were left as they had been originally engraved. Figures 3.11–3.13 illustrate this stage of the process.

Fig. 3.13. J. Stanley Haas building a wax plate. Courtesy Haas Wax-Engraving Co.

Flaming, fusing, or *singeing* was the process of smoothing the form of the molds and flowing the building wax right up to the engraved lines to strengthen the electrotype.[49] It was done with a *flaming tool,* which consisted simply of a short length of copper piping, narrowed at one end, and attached to a gas outlet, which maintained a thin flame about three or four inches long. This flame was brushed lightly over the wax, melting it momentarily, but long enough to smooth it out and run it right up to the engraving. Surface tension usually prevented it from flowing into the lines. (See fig. 3.14.)

Checking

At this stage in the process, a master engraver examined the wax mold and removed any stray specks of wax with a pointed tool, a procedure called *picking up.* Occasionally, slight moisture in the building sticks caused the wax to spatter on contact with the building iron, leaving minute dots of wax in the lines or letters,

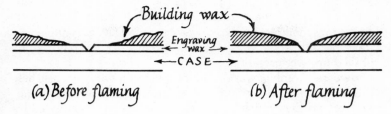

(a) *Before flaming*   (b) *After flaming*

Fig. 3.14.   Flaming. Diagram illustrating the shape of a wax mold before and after flaming

which would, of course, if not removed, show up in the final printing plate.[50] The engraving was also carefully checked to make sure that the silver nitrate coating on the case was intact. Extra silver nitrate was applied with a fine brush to cover the scratched portions.

Electrotyping

The method generally used for making a relief-printing plate from a wax mold was *electrotyping*. This is an electrolytic method of producing a thin shell by the precipitation of copper, nickel, or other durable metal on a mold for the purpose of making a printing plate. It differs from stereotyping, in which molten metal is poured into the mold, which clearly cannot be of wax.

The electrotype process produced extremely precise, strong printing plates from which multiple electrotypes or stereotypes could be cast. The original plate was called a *key* or *master* plate and was kept aside and not used for printing; casts made from the master plate were used on the presses. When these printing plates showed signs of wear, they could be replaced easily with additional casts from the master plate. Electrotypers were in no sense wax engravers, and although the two were mutually dependent, wax engraving was usually done in a separate establishment, or at least in a separate division of the electrotype company.

Graphiting

Electrotyping required an electrically conductive mold to attract ions from the electrolyte, a requirement obviously not satisfied by wax. Any fine-grained electrically conductive substance sufficed, but certain materials were particularly recommended. Gold or silver leaf and nitrate of silver were listed by Thompson, but the most common material was graphite. The graphite could either be applied dry as a fine powder (dry graphiting) or sprayed on (wet graphiting).[51]

Powdered graphite was dusted on the mold with soft brushes. At Rand McNally the graphite was applied with "a row of long, very soft brushes which bounced up and down over the entire surface of the plate." All surplus powder was removed with a blast of air or by flooding the surface with alcohol. The latter not only removed excess powder, but also prevented the formation of air bubbles in the engraved lines.[52]

Wet graphiting was said in 1918 to be "universally recognized as the best method of graphiting the surface of the mold." More recently, the use of silver nitrate, a rediscovery of Thompson's findings of 1845, was found preferable. Both wet graphite and silver nitrate were sprayed on the mold with equipment such as that illustrated in figure 3.15.[53]

A thin layer of copper was applied chemically to the graphited mold to form a base for the subsequent electrodeposition. A solution of copper sulfate (blue vitriol) in the proportions of two pounds of copper sulfate to one gallon of water, was poured or brushed on the wax mold, and powdered iron was dusted into the solution, often from a pepper shaker. The combination of copper sulfate and water produced sulfuric acid and copper, but the presence of iron attracted the acid, which released the copper to be deposited on the mold. Prior to this stage, the back and edges of the plate were insulated against unwanted electrodeposits. To this end, these areas were usually painted with melted wax.[54]

The electrodepositing tank consisted of a heavy lead-lined tank filled with a solution of dilute sulfuric acid (one and a half pounds of copper sulfate to one gallon of water) which was kept agitated throughout the process, usually by passing air through it. For nickeltyping, a solution of nickel salts served as the electrolyte. The anodes, or positive electrodes, were bars of pure copper for copper electrotyping, or nickel for nickeltyping. They were suspended from bars laid over the electrodepositing tank.

Fig. 3.15. Spraying equipment for the application of wet graphite or silver nitrate to the mold. Courtesy A. R. Koehler Electrotype Co.

The negative electrode consisted of the copper case with the graphited wax mold, which was hung on another rod over the tank so that the mold faced one of the anodes. The mold was held by grippers, which penetrated the wax to form an electrical contact with the copper beneath. (See figs. 3.16 and 3.17.)

The positive and negative electrodes were connected to a DC power source of up to 16 volts. Until 1872, when the plating dynamo was invented by Leslie, Smee batteries were used for this purpose. The dynamo reduced the depositing time considerably.[55] The electrolyte of copper sulfate solution broke down into two portions when electricity was passed through the system. The sulfate portion, which was electrically negative, was attracted to the positive copper anode to combine with the copper and form more copper sulfate. On the other hand, the copper portion, being positive, was attracted to the negative electrode, the case and mold, and a copper shell was slowly built on the coated wax. The required thickness of shell ranged

Fig. 3.16. Electrotyping tank. The three electrodes in the foreground are anodes. The electrode identified by the white card is the cathode (the wax mold to be electrotyped). Courtesy A. R. Koehler Electrotype Co.

Fig. 3.17. Grippers. Note the group of cathodes with butterfly screw attachments to the right on the top row. Courtesy A. R. Koehler Electrotype Co.

between 0.016 and 0.006 inch.[56] In the days of the Smee battery, thirty or forty hours of electrodeposition were necessary to achieve this thickness; the dynamo reduced this time to two hours. Meldau reports a time of one half hour, depending on the thickness desired. Higher currents obviously reduced the time required, but great care was needed to avoid over-depositing the shell. The coating on the built-up portions of the mold was somewhat thinner than that on the engraving.

When sufficient copper had been deposited on the mold, the latter was removed from the electrotyping bath. The copper shell was then hosed down with boiling water, which melted the wax mold. The copper shell could in this manner be easily separated from the wax. If the shell was damaged in any way at this stage, the work usually had to be done again from the beginning. To lose an engraving in this way was one of the most frustrating moments for the engraver.[57]

It was necessary to strengthen the thin copper shell by a process called *backing up* in order for it to withstand the demands of long runs on the press. The copper shell, which was already fairly rigid, was laid face down on a flat surface, and the inside of the shell was wetted with soldering flux and covered with thin sheets of solder. Type metal was then poured in to a depth of about a quarter of an inch, and this bonded to the copper with the solder. It was at this stage that difficulties were encountered in maintaining dimensional stability.[58] The boiling water and then the molten type metal expanded the copper shell to such a degree that adjustments usually had to be made at a later stage to maintain correct registration. Such adjustments often involved cutting and splicing the printing plates.

The preparation of the electrotype shell for printing, known as *electrotype finishing*, was an extremely important stage of the process. Upon the skills of highly specialized electrotype finishers depended the final printing quality of the plate. If the plate was not perfectly flat, an expensive procedure known as make-ready was needed. Here, the pressman pasted small pieces of paper to the bottom of low parts of the plate to compensate for the unevenness in the printing surface. In the waning stages of wax engraving in the early 1950s, when old wax-engraved plates were being corrected but no new work was being done, the finishers were the linchpin of the whole operation.[59]

Corrections were often made on the electrotype of the key or master plate, from which the press plates were cast. Russell Voisin reports that as many as seventy-five separate corrections occurred in one plate. Electrotype corrections were practically impossible to detect in the finished work. Where several changes in line work and lettering had to be made, such as in the inclusion of a new dam and reservoir, that part of the electrotype was cut out with a coping saw and patched with a new wax-engraved electrotype or even a photoengraved zinc plate of the area, which was soldered in place from the back.[60] To repair a depression in the printing surface, the finisher dug into the electrotype on both sides of the depression, pushed up the metal to form a bump, and filed this down to the right thickness and height. Small flaws could be engraved out with ordinary wax-engraver's tools. Where one word or letter required correction, a

Fig. 3.18. Electrotype finisher's bench at the A. R. Koehler Electrotype Co. Note the steel brush for cleaning the highly polished steel plate against which the electrotype was leveled. Courtesy A. R. Koehler Electrotype Co.

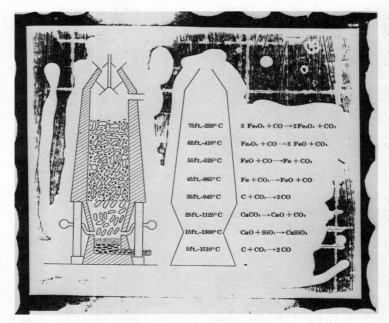

Depth-Temp	Reaction
75ft.-230° C	$3\ Fe_2O_3 + CO \rightarrow 2\ Fe_3O_4 + CO_2$
65ft.-410° C	$Fe_3O_4 + CO \rightarrow 3\ FeO + CO_2$
55ft.-525° C	$FeO + CO \rightarrow Fe + CO_2$
45ft.-865° C	$Fe + CO_2 \rightarrow FeO + CO$
35ft.-945° C	$C + CO_2 \rightarrow 2\ CO$
25ft.-1125° C	$CaCO_3 \rightarrow CaO + CO_2$
15ft.-1300° C	$CaO + SiO_2 \rightarrow CaSiO_3$
5ft.-1510° C	$C + CO_2 \rightarrow 2\ CO$

Fig. 3.19. Proof of an electrotype plate before routing. Note the impression of the bearers around the margins of the diagram. Courtesy Wallace B. Mitchell

hole or slot was drilled in the electrotype, and actual foundry type was plugged in, soldered in place, and the balance of the type stem sawn off.[61]

The printing surface was leveled by laying it face down on a very smooth surface and hammering lightly on the back with a wooden block and mallet. (See fig. 3.18.) To avoid damage to the detail at its edges, the plate was surrounded with *bearers*— flat rims about a quarter of an inch wide which served to support the remainder of the plate. These rims were usually engraved in the original wax plate with a bearer tool. The bearers printed as thick black lines in the proof stage (fig. 3.19) but were removed before final printing by *routing.*

Routing was necessary to remove unwanted parts of the electrotype which might print when inked. This stage was done with a routing machine, consisting essentially of a rotating drill bit which could be moved over the entire surface of the plate as well as vertically.

The finished electrotype, eleven points thick, corrected, leveled, and routed, was finally mounted on a stable base to bring it to type height. The base was usually of hardwood, such as cherry, but occasionally type metal was used. The electrotype was either nailed or soldered to the base, depending on whether wood or metal was used. A large wall map could consist of as many as 125 separate pieces nailed or screwed to the base.[62]

Electrotypes or stereotypes were cast from the master plate, finished, mounted, and beveled at the edges to fit the locking clamps on the bed of the press. Wax engraving produced letterpress plates, and these were printed in essentially the same manner as foundry type. The plate was cleaned and inked, and press proofs were taken until the correct impression pressure was achieved.

In color work, a series of plates, one for each color, was required. The first plate, the key or master plate, contained all the detail that was to appear in black. Proofs taken from this plate were then offset to produce reverse images. While still wet, these reversed proofs were then laid face down on the wax molds for each color, and the image was transferred to the surface of the wax by burnishing the back of the proof. Separate color plates could then be engraved. Halftone blocks were commonly used in conjunction with wax-engraved plates in map work.[63]

4 The Technicians

The need for cooperation between the artist and the printer has been well summarized by R. B. Fishenden in 1929, and his comments are applicable to the connections between cartographers, engravers, printers, and publishers during the period of wax engraving:

> Before the introduction of photographic methods of illustration, the artist himself frequently cut or engraved his metal plate or wood block, and consequently secured harmony in his prints. Both wood-cutting and wood-engraving have their limits of expression, as has also lithography; and the beauty of an etching, aquatint, or mezzotint arises from the recognition by the engraver of the qualities and limitations of the printing medium.
>
> Unfortunately the communion between the artist and the printer decreases as the industry develops, and in place of the artist being a printer or working in close touch with one, a new organization (i.e., relationship) has grown providing little or no connexion between the designer and the printing press.[1]

Similarly, the connections between the various personnel engaged in the production of a map have been noticeably weakened since the introduction of photomechanical techniques in the latter half of the nineteenth century. Before this time, in the eras of woodcut and copperplate engraving, the cartographer often engraved and printed the map himself, or at least had the facilities at hand in his own establishment. Some of the renowned map publishers of the Netherlands in the sixteenth and seventeenth centuries, for example, Gerhard Mercator, Abraham Ortelius, Jodocus Hondius, and Willem Janszoon Blaeu were all, at one time or another, cartographers, engravers, printers, and publishers, and consequently possessed an intimate knowledge of all stages of map production. Whether they themselves did the work or supervised it, there appeared from their publishing houses a product which reflected a harmony between cartographer and printer.[2]

With the introduction of the power press, photomechanical engraving, and other mechanical aids to the graphic arts, each stage of the mapmaking process became increasingly complicated, and each man in the chain had to become increasingly specialized to master the new technology. In other words, the production line had arrived in cartography. It was now not usual for one man to become a competent cartographic compiler, engraver, and printer as in the copperplate era. Each man now had a special skill. As one might expect, a breakdown in communication among the specialists resulted.

Wax engraving provides a good example of production-line techniques in cartography and map reproduction. Two features of the technique are especially interesting. In the first place, wax engraving was the first major map-printing technique to begin in the increasingly industrialized nineteenth-century era of printing; and second, the technique encouraged a particularly large number of specialist categories of personnel.

There are probably two reasons why wax engraving was the first production-line technique for the making of maps. First, the technique employed a relief-printing plate which could take advantage of the cylinder power-press which had been developed sufficiently for commercial use by 1839 when wax engraving was introduced, while copper engraving and lithography could not be adapted to such developments at that time. Second, since the mechanical basis of wax engraving was electrotyping, which had been introduced in the 1830s and had become commercially workable by the following decade, the successful development of the technique in no way depended on improvements in photography or photomechanical techniques, despite the occasional refinements which photography added to some stages of the process, such as in the transfer of images to the wax. Wax engraving was consequently well established by the time the photographic adaptations of lithography and intaglio engraving had become commercially feasible in the closing decades of the nineteenth century.

The large number of specialist categories that developed in the engraving and platemaking stages of the wax-engraving process included stampers, compositors to set the type for the stampers, line engravers, builders and pickers at the engraving stage, and electrotypers and electrotype finishers at the platemaking stage. Contrast this with the case of direct or transfer lithography, which employed usually one lithographic draftsman and one lithographer or lithographic transfer man. The extra stages required in wax engraving stemmed largely from the use of printer's type for the lettering and the two-stage electrotyping procedure.

The Engraver

On coming into the wax-engraving business, a young man served an apprenticeship which lasted about two years. The most menial tasks, done by the least-experienced apprentices, were running errands and preparing plate for the line engravers by scrubbing them with pumice and flowing on a layer of wax. They were introduced to the job of building after two weeks or more, depending on their promise. Only coarse work was done before handing the plate to the experienced builder. This gave the apprentice good practice in controlling a tool and getting used to the consistency of the wax.

Before the apprentice was given a job of freehand engraving, he was set to work ruling tints on the wax plate with a ruling machine. This work gave the apprentice useful experience in feeling the response of wax to the engraving tool, yet was a relatively straightforward task that could be learned in a short time.

The goal of the apprentice was usually to become either an accomplished engraver or stamper. Occasionally, a craftsman would do both engraving and stamping, but this was not the rule. The engraver was entrusted with the task of engraving all line work, while the stamper pressed in the type for the lettering and the symbol dies for town dots and other features. They required different skills and appealed to different people. One engraver interviewed by the author said straightforwardly that he did not like stamping and was not good at it. When his line work was done, therefore, he would hand the plate to a specialized stamper for the lettering.[3]

While the engravers and stampers were skilled craftsmen, their training in cartography was often severely limited. Consequently, their roles as cartographers in the production of the map were variable and complex, meriting separate discussion here. In the large map-printing companies such as C. S. Hammond, Rand McNally, and Matthews-Northrup, the work of compilation was overseen by chief cartographers who had at least some geographical background. Often these master cartographers served in an advisory capacity to the company, and while they were on the payroll, held positions elsewhere, frequently in the universities. The best known of these in the United States was J. Paul Goode, who while a professor at the University of Chicago advised for Rand McNally, giving his name to their *Goode's School Atlas*. During the 1920s, when this atlas was published, Rand McNally also had three other chief cartographers in their employ, who also worked in an executive capacity.[4] Beneath them were the cartographic editors, who prepared the original compilations for the engraver. In the case of Rand McNally, the engraving was done within the firm. The company offered a training class in wax engraving, with as many as twelve persons in a class at a salary of three dollars a week.[5] The engravers worked from a detailed, well-drawn compilation photomechanically reproduced on the surface of the wax, and consequently had no design or editorial responsibilities.

At C. S. Hammond the procedure was slightly different, in that the engraving was not done by the firm itself but contracted out to engraving and electrotyping firms. The method of compilation, however, was similar. The company also had their cartographic advisors, and the services of Armin K. Lobeck and

Erwin Raisz were frequently enlisted. According to Ferdinand von Schwedler, for a long time a cartographer at C. S. Hammond, the procedure for making a new map was to have an editorial board meeting, with the outside advisor present, in order to discuss what should be shown on the map. The compilers or staff cartographers then produced drawings as copy for the engravers. It is significant that most of the firm's cartographers were European with geographic or cartographic training. German was spoken in the drafting room. Von Schwedler's geography was gained at the University of Berlin as part of a military training. As a lithographer on stone with the German General Staff in 1916, he had never heard of wax engraving. When he emigrated to the United States in 1920, he learned wax engraving for engraving small maps in books at A. J. Nystrom.[6] This is another indication that, for the most part, wax engraving was limited to the United States.

The compilation drawing was then sent by Hammond to a large engraving firm in New York, such as L. L. Poates or the Empire Wax-Engraving Company, where the wax engraving was done and the printing plate made. In such a situation, the engraver had no responsibilities for the content of the map or the line weights or type styles and sizes, as this work had been done by the Hammond cartographers entrusted with the design of the map.

While the engraver did not have much say in the design or editing of a map published by the large map companies, there were a great many smaller wax-engraving firms in the country where the engraver frequently assumed the role of cartographer. It is interesting to see how in these cases the making of maps found its way into what could be classified as printing establishments, which was in contrast to the case of the cartographic draftsman who prepared pen-and-ink drawings for the camera. To use an analogy from economics, pen-and-ink drafting was light industry requiring little capital equipment and hence of a "footloose" character, not tied to any other industry for services. On the other hand, wax engraving could be compared to heavy industry, since it required capital equipment and was tied to engraving firms or electrotyping plants.

The capital equipment of wax engraving included, first of all, a considerable stock of printer's foundry type and the equipment to set it and pull proofs. In addition, the engraver required ruling machines to rule tints, and often a stamping machine to press the type into the wax. His shop had to be equipped with gas jets to feed the building irons and flamers. Finally, since the wax plates he produced were both bulky and fragile, the electrotyper had to be close at hand. The combination of these factors created a close association between the wax engraver, electrotyper, and printer, so that time and again we find a wax-engraving shop as a department of an electrotype foundry, or at least very close to one (see figs. 4.1 and 4.2).

Thus, the map was often considered by both client and engraver simply as a piece of printing. Since it consisted of a combination of lines and letters, it could be classified with the ruled form and the engraved dial, excellent subjects for the wax engraver. Other than being a combination of lines and lettering, it was not considered a special problem. The wax engravers who handled maps consequently had to pick up whatever cartographic knowledge they needed as they went along. Such was the experience of Stanley Larson:

> There was no effort made for formal training [in cartography] whatsoever. And when I came in, I was only a kid, and the first maps I did were some maps of West Virginia, mostly black and white, but one big color map in there. I was given that job after about a year. If anyone knew the amount of knowledge that went into it, they would probably be startled and alarmed. I didn't have that much cartographic knowledge. For things like projections, I went to books; even today, my knowledge of projections is so limited. I have to keep refreshing my memory all the time when something comes in. The apprenticeship didn't have any cartography. It wasn't very organized or formal at all.[7]

A specific example of what was entrusted to the engraver was revealed in an interview with another wax engraver, Wallace B. Mitchell: "The engraver was to some extent a cartographer. The editor would tell you how many names to put on a map, but this was often a vague indication, such as 'all names referred to in the

Fig. 4.1. Industrial building in Buffalo, N.Y., demolished in 1969, which contained three engraving firms, one electrotype plant, and various other enterprises relating to the printing trade

text' for Biblical maps. Type styles and line weights were the engraver's decision."[8]

The engravers just quoted were able to achieve good results despite a lack of training, as a result of a certain amount of natural ability, common sense, and artistic talent. Unfortunately the same cannot be said of the engravers of all firms, as illustrated by the following anecdote by von Schwedler:

> There was a story about the Geographical Publishing Company in Chicago. They hired someone to make them a globe. This was something new. So he bought a globe from Rand McNally or some other company and put it in the bath tub until he had all the globe gores dissolved. Then he pasted them up flat on boards, one for the northern hemisphere and one for the southern hemisphere. Only he

Fig. 4.2.   Name board on the building illustrated in figure 4.1

forgot he had to paste them up in reverse for the southern hemisphere. He did it the same way. Then he had this thing photographed at a camera place in Chicago, and I just happened to be there and saw this. And I said to the camera man, "that don't look right" and he said to me, "look here, this is another customer, don't you mention anything about it." So this poor guy printed five thousand globes there, and he had to throw them all away because he had South America to the east of Africa, and so on. That's one thing that can happen when a man is technically a draftsman but doesn't know any cartography.

Von Schwedler also pointed out the lack of cartographic talent of the supervisors of the engravers: "Our bosses were salesmen. They were on the road selling stuff. They haven't got heart or sense to know how its done. They don't care. As long as they get it done."[9]

This is yet another indication that there was a distinct business orientation to wax engraving, with perhaps a tendency to regard quality as secondary. As long as a good deal of mapmaking was in the hands of wax engravers, the map was essentially something to be taken to one who had been trained as an apprentice in the printing trade and not as a professional cartographer. (See fig. 4.3.)

We would expect to find such limitations manifested most clearly in the engraver's attempts at generalization, as this was an aspect of cartography which was frequently left entirely to his care. One such aspect of generalization was in the selection of information, a procedure that, if it is to be done well, requires a great deal of cartographic experience. Figure 4.4 is an example of poor selection procedure.

Another characteristic of the typical wax-engraved map which revealed the cartographic aptitude of the engraver was the placement of type. Often no attempt was made to position names exactly at the compilation stage, since the exact size of the type was difficult to judge, and the engraver or stamper consequently was given freedom to move the names around. This frequently resulted in cramped and awkward arrangements.

Fig. 4.3. Views of two wax-engraving shops in the 1930s. *Top*, Back Bay Electrotype Co., Boston, 1937. *Bottom*, Magnuson and Vincent, 1937 (note the stamping machine in the foreground). Courtesy Wallace B. Mitchell

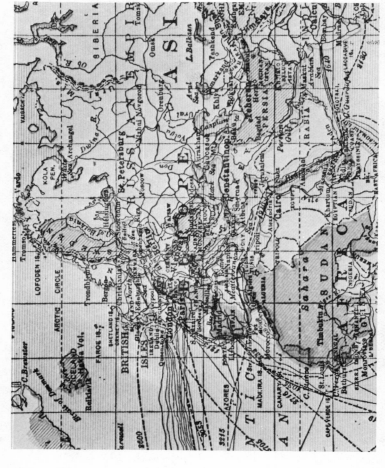

Fig. 4.4. Detail of wax-engraved map illustrating several generalization problems. From *Hammond's Pocket Atlas of the World* (New York: C. S. Hammond and Co., 1913), p. 3

The Technicians

THE ELECTROTYPER

The two main groups of personnel engaged in the platemaking stage of wax engraving were the electrotypers and the electrotype finishers. The role of the electrotyper was to cast the relief-printing plate from the engraved wax mold, and he was therefore entrusted with maintaining the quality of the engraving. Two stages were particularly critical in his work: electrodeposition and separating the plate from the mold. First, if he were to leave the plate in the electrotyping tank too long, the line weights of the engraving would become progressively heavier, a phenomenon somewhat analogous to overexposure of printing plates in photomechanical work. Second, if the copper shell were in some way to become bonded to the copper case beneath, it usually could not be salvaged when separated, and the engraving would have to be redone from the beginning. While the engraver knew of these dangers, there was little he could do to avert them once his work had been sent to the electrotyping foundry. Besides, the electrotyper was a highly specialized and unionized member of the printing trade and was unlikely to entertain the critical surveillance of one of his clients.

The electrotype finisher was responsible not only for preparing the electrotype plate for printing by routing, leveling, and mounting, but also for the important job of correcting old plates. For this reason, large map companies using wax engraving commonly employed one or more finishers at their plant.[10]

5 Technique and Style

It is important to recognize the distinction between resistance and autography in a medium. An autographic medium is one in which the artist can work directly with the process, bypassing the engraver, thus theoretically enhancing the directness of his expression. It should also be recognized that there are practical advantages to the autographic method that have probably been active in the history of printmaking: it was cheaper, more convenient, and quicker to cut out the middleman.

A resistant medium may be defined as one that places many technical constraints on the engraver, which are in turn reflected in the style of the print. In other words, the technique affects the style *in spite of* the technician. It is resistance, not autography, with which we are concerned here. The woodcut may be cited as being an extremely resistant medium. A woodcut looks like a woodcut. Lithography, on the other hand, has less resistance in the sense that it can be molded by the lithographer, with the result that he can imitate in the medium almost anything from chalk drawings to copper engravings. Wax engraving, as a map reproduction technique, was on the resistant side. A wax engraving looks like a wax engraving. The tools did not allow a great deal of freedom to the engraver since the technique was well adapted to the use of mechanical tools: for example, stamping dies, type, and gravers of a fixed gauge. Hence the technique tended to leave its mark on the styles of the maps it produced.

Point Symbols

To the modern cartographic draftsman, with his electric point scribers and preprinted point symbols, the drafting of consistent, regular points is a straightforward task. To the wax engraver of the nineteenth century, it was even more straightforward. Here was a distinct advantage of wax engraving over its competition, lithography. Regular, consistent point symbols could be produced with symbol dies stamped perpendicularly through the softened wax to the copper case beneath. (Fig. 5.1.) There was no ink to run, smear, or flake off; no struggling with dotting pens. While the lithographic draftsman frequently had to draw small point symbols by hand, or with a compass, the wax engraver could stamp in the appropriate symbol in one easy movement. However, larger point symbols, such as graduated circles, and large pictorial symbols were rarely stamped, but were engraved with compasses or other hand tools, and consequently these do not share the same characteristics of regularity and consistency.

Occasionally the engraver could make his own symbol dies, but there were also some commonly used commercial patterns. As a result, a small number of symbols became standard. One engraver interviewed by the writer identified two of his point symbol dies by what they had come to represent: ⊛ was a state capital, ⊙ was a county seat. These were commercially produced tools that adopted a special meaning for this engraver, and no doubt for others. He spoke as if it were a common convention, and for some years had used the same dies to symbolize state capitals and county seats.[1] Perhaps this affords us a glimpse of the source of a cartographic convention.

To summarize, point symbols produced with stamping dies on wax-engraved maps have a number of characteristics. They are regular in shape, each symbol on any one map being fashioned from the same tool; they are sharp in definition, because of the fine molding quality of beeswax and the resulting sharply defined electrotyped plate; and the symbols tend to conform to

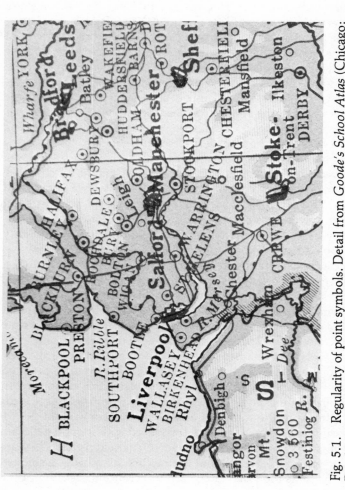

Fig. 5.1. Regularity of point symbols. Detail from *Goode's School Atlas* (Chicago: Rand McNally and Co., 1939), p. 116. Note the regularity of the town symbols in the Manchester area

conventions set by the limited number of symbol patterns available commercially.

LINES

In the discussion of wax-engraved point symbols, some similarities become apparent between wax engraving and scribing on glass or on plastic sheets. Both are engraving techniques, and in the sense that the draftsman or engraver removes material instead of adding it to produce the image, both wax engraving and scribing are essentially the opposite of pen-and-ink drafting. In scribing, a thin scribecoat is removed with an engraving tool; in wax engraving a thin layer of wax is removed in a similar way. It is at the platemaking stage that the similarity breaks down; in scribing, the plates are produced photochemically, while in wax engraving they are cast electrochemically. Nevertheless, modern scribing could be described as a kind of photographic wax engraving, and it is not impossible that the very idea of coating glass with an engravable coating may have had its roots in the practice of coating glass with wax in wax engraving.[2]

It is in the character of some categories of line work that the similarities of scribing and wax engraving are especially striking. The line tools used by the wax engraver, as well as those used by the scriber, were "mechanical" in that they were rarely able to transmit the subtleties of a craftsman's individual style in the same way that is possible with ink drafting tools such as a crowquill pen. In the latter example, the nib flexes under the slightest pressure, yielding a thick or thin line at the draftsman's discretion. The wax-engraver's and scriber's tools, on the other hand, were essentially unsympathetic to differences in pressure imparted by the engraver. A line tool ground to four thousandths of an inch produced a line four thousandths of an inch wide, whether engraved by craftsman A or B, at any time, since the tool had to cut all the way through the wax to the copper case beneath, and could accommodate no difference in pressure. In this, the technique differed markedly from the copper or steel intaglio techniques, in which the engraver could vary the width of his line at will by varying the pressure of his palm on the burin.

Chapter Five

Ruled Lines

Fine hairlines, engraved with needlelike tools along a straight or french curve, or by a ruling machine, are characteristic of many wax-engraved maps. Line thicknesses of .002 inch were quite feasible at the engraving stage.

Thicker lines, such as those employed for map borders, were usually engraved by V-shaped tools mechanically ground to a set width, usually according to some standard scale based on thousandths of an inch or on the printer's measure called *point* (approximately 1/72 inch). When precisely ground, such tools conformed to a hierarchy of widths, encouraging consistency of line thickness from map to map. For example, while the lithographic draftsman might have great difficulty in repeating exact line widths when using a ruling pen for the borders of different maps in an atlas, the wax engraver could simply select the tool assigned for the borders, thus ensuring consistency.

Occasionally, thicker lines were built up by ruling a number of hairlines adjacent to one another, as in figure 5.2. Although this facilitated the engraving of extremely neat and sharp corners, the method had little else to commend it, especially as it sometimes resulted in a gray line when the lines were not drawn close enough together.

Freehand Lines

Single, freehand lines, such as those used to represent rivers, coastlines, lakes, boundaries, and so forth, were engraved with two main types of tools. The stylus, a pointed, needlelike tool, was used in the hand like a pencil to draw through the wax. This type of tool was characteristically used by Palmer and Dawson to engrave line drawings in wax in the nineteenth century, but its use has often been extended to the engraving of maps.

The other main type of tool was V-shaped, and this was pushed through the wax like a graver, producing a clean, regular line. These tools were also ground to standard widths, and hence conformed to a standard hierarchy, consistent from map to map. The engraving of rivers posed special problems; they often required to be tapered off toward their source. The wax engraver met the problem in the same way the scribing draftsman meets it

6

Fig. 5.2. Map border consisting of multiple lines. Courtesy Wallace B. Mitchell

now. Different tools were used for different sections of the river, smoothing over the breaks with a needle or other fine tool.

Hand-engraved lines in wax engraving exhibit what may be called particular sinuosity characteristics in their style of generalization, arising both from the type of engraving tool and the direction in which it is worked. For background, it might be helpful to summarize the various sinuosity characteristics associated with four main types of tools used in map engraving at various stages in the history of map printing: the woodcutter's knife, the graver or burin, the stylus, and the swivel scriber.

Imagine that an engraver is directed to copy the sinuous line shown in figure 5.3 with the four engraving tools in question. The lines illustrated are intended to represent the lines on the block or plate. Each tool will contribute its own character to the resulting line, both in terms of angularity and regularity. We will assume that (*a*) is perfectly smooth and regular.

The woodcutter's flat-bladed knife will tend to produce a characteristically angular line, often of uneven thickness, as in (*b*). The knife is difficult to control against the resistant hardwood, resulting in frequent slips and irregularities. On the other

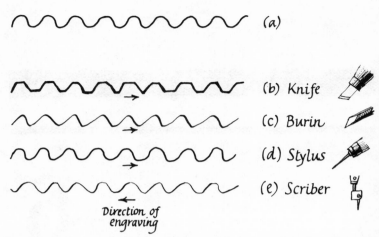

Fig. 5.3. Sinuosity characteristics of lines produced with various engraving tools

hand, the burin or graver used by copper, steel, or wood engravers, held firmly against the palm of the hand and pushed away from the operator, is capable of a flowing, elegant, delicate line unmatched by any other engraving tool. But, as a result of the pushing motion, it is natural to both "overshoot" and engrave more deeply in one direction than another, which produces a slight skewness and a "thick-and-thin" character in the line (*c*).

A pointed tool, such as the etcher's stylus, produces characteristically rounded curves (*d*) which probably approximate the character of (*a*) more closely than the other tools described here. Held in the hand like a pencil, it allows considerable control, especially when engraving in material of little resistance, such as soft wax. Since the cutting part of the tool is round, the direction of the line has little bearing on the resultant thickness. Consequently, etched lines are frequently both rounded and regular.

The swivel scriber represents the most modern of the tools described here. The cutting edge of this tool is automatically maintained at a 90 degree angle to the line being engraved, because the point is allowed to swivel freely as it is pulled toward the operator. There is, however, a short lag which often

100

results from this pulling motion; in other words, the tool may not change direction precisely when the operator requires it, but slightly afterward. The experienced scribing draftsman obviously compensates for this, and it would be an exaggeration to suggest that all scribed curved lines invariably show a skewness as a result of the lag, but there is a tendency for the tool to behave this way.

The wax engraver used both the burin-type tool and the pointed stylus in his hand-engraved line work, and it is sometimes possible to recognize which type of tool was used by examining the line characteristics.

A hand-engraved wax-engraved line produced with the burin is similar to that produced by the copperplate engraver, except that the wax was less resistant than copper or steel, and the angularity was therefore not as pronounced. The line produced in wax with the stylus, on the other hand, will approximate that of the etching, that is, rather more rounded and regular.

Double or triple lines, often used to symbolize roads, railroads, canals, and so forth, were easily engraved in wax by using specially made tools, such as the double-line graver. It requires great skill to be able to draw two curved lines parallel to each other with a single-line tool, and the two-pronged graver effectively eliminated this problem. Here again is common ground between wax engraving and scribing: the ease of engraving multiple lines. Figure 5.4 shows some good examples of the regularity and consistency possible with the multiple-pronged tool.

Stamped Lines

Composite lines, or lines composed of other symbols, such as dashes, dots, crosses, and so forth, are usually devised to enable the reader to recognize a large number of line categories when no hierarchy is implied. The wax engraver often engraved these lines with wheellike tools having teeth cut into the edge of the wheel which would create the appropriate symbols (fig. 5.5). It was difficult to use these tools on extremely intricate lines, for which freehand tools were used, and that they reserved for composite lines of straight or simple configuration. For example,

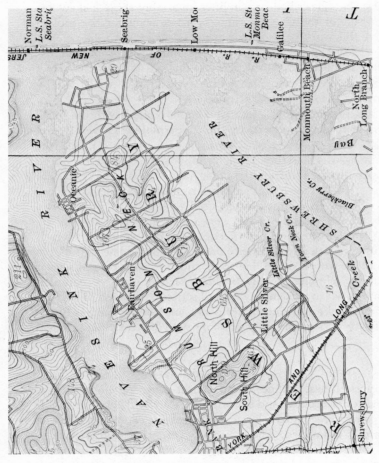

Fig. 5.4. Detail of USGS map. From *Poates Wax Engraving Superiority* (New York: L. L. Poates Engraving Co., 1913), p. 6

Fig. 5.5. Wheellike tools for engraving dotted or diced lines. Courtesy J. Stanley Haas. Photo, Peter Weil

the dashed or diced line wheel tool could be used with advantage for ocean routes.

Essentially, the wheellike tool produced a series of point symbols pressed into the wax. Consequently, lines made from these tools share the characteristics of point symbols made with symbol dies: regularity, sharpness, and consistency.

In addition to the influence of tools on the various characteristics of lines in wax engraving, some other, less significant

103

factors were sometimes responsible for modifying these characteristics. For example, if the engraving wax had not been ground fine enough, it would exhibit a powdery consistency, and small pieces would chip out when engraved, leaving a ragged edge to the line.[3] Another technical flaw which affected the final printed line occurred when the engraving tool cut into the copper case by accident, causing a small bump on the electrotype which would print darker than the surrounding part of the plate.[4]

<h2 align="center">Area Symbols</h2>
<h3 align="center">Flat Tints</h3>

Flat tints composed of lines (ruled tints) were commonly engraved on a precise ruling machine, such as that previously described. The intention was to create a flat tone in which the individual lines would not be visible to the naked eye. In wax engraving, however, flat tints were likely to have two special characteristics. In the first place, the lines could not ordinarily be ruled as closely together in wax as on a lithographic stone, because the wax coating between the lines tended to disintegrate when the ruling was too fine. Secondly, a wide band was frequently used to fringe the ruled area. In addition to serving the obvious visual function of emphasizing the shoreline, border, or boundary on a map, the wide band also served as an excellent device for neatly finishing off the edges of the ruling. The engraver simply ran the wide tool along the ends of the ruled lines. The resulting wide line also served as a useful *bearer* against which the plate could be leveled, and the bearer protected the fragile, thin ends of the ruled lines from being damaged when the plate was being moved or printed.

Flat tints in wax engraving could also be composed of small points, obtained either by pressing a stylus into the wax by hand, or by ruling the plate with a special dotted-line tool, producing the patterns illustrated in figure 5.6. Both of these methods were also applied to copper or steel engraving, when the points were punched or ruled into the plate by means of a variety of tools. Solid tones were readily produced in wax engraving by the method described earlier.[5]

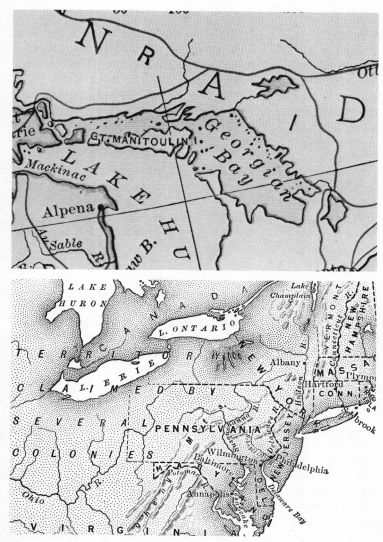

Fig. 5.6.   Stippling by engraving machine (top) and by hand (bottom)

Chapter Five

Graded Tints

Although graded hand-stippling was quite possible in wax engraving, it would have been an extremely slow and laborious method of obtaining a continuous graded tone. The standard method of achieving this end was to use a halftone screen. As it was not possible to produce a halftone plate by the wax-engraving technique (which is essentially nonphotographic), such plates had to be prepared and printed separately. This was expensive and reserved only for the highest quality work (see fig. 5.7).

The general effect of this inability to produce satisfactory halftones in wax engraving was to discourage the representation of any feature with a graded tone. Thus the representation of relief often relied on symbols such as contours or hachures. Punches were developed to produce mechanical-looking hachures (see fig. 5.8).

LETTERING

Of all visual characteristics of small-scale atlas maps, lettering holds perhaps the dominant position. It is probably the first element of the map to draw comment from a critical viewer, unless the lettering is so good that, as in Beatrice Warde's analogy, it becomes "invisible."[6] In addition, the lettering of maps has consistently caused problems for the map printer, and the way in which these problems were overcome often affords the first clue to the kinds of engraving tools used or the printing methods employed. Distinctive are the fine, delicate swash letters made by the graver in copper and steel, the rounded letters created by the etcher's stylus, or the angular work of the woodcut.

A fundamental distinction may be made between hand lettering and lettering produced by printer's metal type. A trained eye can usually discern the differences immediately. Hand lettering is characteristically freer, less mechanical looking, and hence, in the eyes of some observers, more artistic. Type, on the other hand, usually appears regular, stiff, mechanical, and is therefore often considered lacking in artistic merit. It should be pointed out, however, that badly produced hand lettering may be

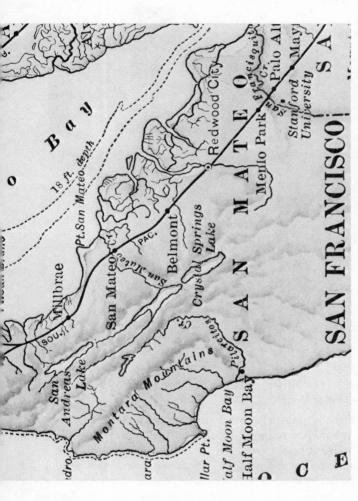

Fig. 5.7. Combination of wax engraving and halftone block. From *Poates Wax Engraving Superiority* (New York: L. L. Poates Engraving Co., 1913), p. 5

Fig. 5.8. Hachure punches. Courtesy Rand McNally and Co. Photo, Peter Weil

exceptionally unartistic, while thoughtfully chosen and applied type on a map may be no less beautiful than its equivalent in a finely printed book.

Hand lettering was possible in wax and indeed was probably a widespread method in the early days of wax engraving, as in the maps of Sidney Edwards Morse from 1839 to 1845. Indeed, it was probably easier to hand letter in wax than on a lithographic stone or copperplate, because the wax engraver could perform the work right-reading as it was to appear in the final map, while lettering on the lithographic stone or engraved into the intaglio copperplate had to be reversed, unless, of course, it had been drawn on transfer paper and then floated off the paper onto the stone.

In the later years of the technique, however, the use of printer's type for the lettering on wax-engraved maps became a standard procedure and a striking characteristic of these maps. While the use of movable type was not unknown in lithography, as strikingly illustrated by Eckstein's method,[7] the vast majority of the lettering on lithographed maps was hand produced until the invention of stick-up lettering. Technically speaking, wax engraving was extremely well adapted to the use of letterpress type; brand new foundry type could be pressed through the wax to make perfect letters. From the personnel aspect, the method obviated the need to search for artistic cartographers who could letter well by hand, a rare breed in the American map printing world of the late nineteenth and early twentieth centuries.

This dominant use of type in wax engraving gave rise to several characteristics. Of course, a general characteristic already alluded to is that wax-engraved maps are "typey," to borrow a most expressive term used by some young subjects who took part in one of Barbara Bartz Petchenik's studies on cartographic legibility.[8] But we can break down "typey" into a number of component characteristics relating to the amount of type, its placement, size, and style, as well as to some technical oddities which sometimes attend its use.

The amount or density of type on wax-engraved maps obviously varied considerably according to the policy of each map compiler as well as to the area covered, the purpose of the map, and a host of other factors. Examples can be found of both extremely crowded and extremely sparse lettering, even from the same map publisher. Generally speaking, however, the wax-engraved map erred on the side of too many names rather than too few. To the clientele of such products, the usefulness of a map was often measured by the number of names appearing on it (which should also include the reader's home town, regardless of size), and wax engraving accommodated such desires admirably. A skilled stamper could produce a large number of small, clear names extremely efficiently. Thus, the technique certainly did nothing to hinder overcrowding, and may in some cases be claimed to have directly encouraged it. Nevertheless, we should avoid a direct and simple deterministic relationship between the use of wax engraving and the overcrowding of a map with names, since the role of the clientele was probably equally important.

The use of type on maps usually poses certain positioning problems that are not faced by the more versatile hand lettering. These problems were rarely overcome by the wax engraver and result in a mechanical awkwardness giving rise to the tendency to fit names at all angles in small spaces, and to split words to avoid line work.

The wax engraver usually employed straight or curved stamping sticks or type holders for each name on a map. It was much easier to fit a straight name at an angle in a small space rather than select a curved stamping stick of appropriate radius to fit the name in the available space. Straight names at an angle have a decidedly awkward appearance, especially when neighboring names are also straight and at different angles (see fig. 5.9). It has now become an accepted convention in type placement on maps to curve any name which, for reasons of space, is required to deviate from the normal orientation parallel to the lines of latitude.[9] Frequent violations of this principle may be observed on wax-engraved maps.

The difficulty of fitting type into a small space sometimes led to the arbitrary splitting of names where they would otherwise encroach on line or other detail (see fig. 5.10). A contributing factor to this problem was the order in which the line work and lettering were engraved. Some engravers preferred to engrave the line work before the lettering was stamped in, thus encouraging the splitting of words rather than breaking the line work.[10]

Since the lettering on wax-engraved maps was charactertistically produced with type, it conforms to a hierarchy of sizes dictated by point size. The Hammond type index reproduced as figure 3.6 shows the point sizes available to wax engravers in the 1930s and 40s. The smallest size according to this index appears to be 2 1/2-point Gothic 545; the smallest the author has seen used on maps is 3-point Century Schoolbook, which could be considered sufficiently small for any cartographic purpose.[11]

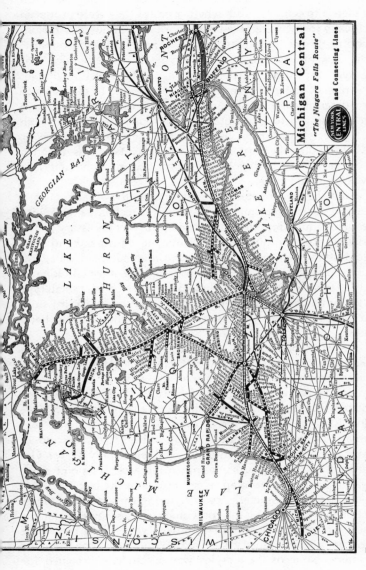

Fig. 5.9. Straight, angled lettering. From *Atlas of Traffic Maps* (Chicago: LaSalle Extension University, 1930), p. 148

It is thought that such small sizes of type were not cast until the nineteenth century,[12] but whether or not they were produced specifically for wax-engraved map work is not known. Hodson, describing the methods of engraving "typographic etched maps" by the Dawson process of wax engraving, is of the opinion that they were cast for this purpose: "So efficient, as well as simple, is this plan of lettering, which is also applicable to diagrams of every description, that the inventors (Messrs. Dawson) have been encouraged to types of an exceptionally small, although still legible, character cut and cast for their exclusive use."[13] Notwithstanding the purposes for which such type was designed and cut, however, the use of minute type became a characteristic of wax-engraved maps.

It should be remembered that no reduction took place between the engraving and platemaking stages of the technique. If 3-point type were required on the finished map, it was necessary at the engraving stage. In photolithography, on the other hand, one could draft the map with hand lettering at an enlarged scale, and reduce it photographically at the platemaking stage. As long as small lettering was the fashion in wax-engraved maps, small type was by far the most efficient method of achieving it, for it would have been impractical to engrave letters of that size by hand in wax.

Larger sizes of lettering conformed to standard point sizes available to the engraver. Often there was a range of intermediate sizes which might surprise the modern printer. The Hammond type index indicates that Gothic 545, apart from being available in 2 1/2, 3, and 4-point sizes, was also on hand in 5, 6, 7, 8, and 10-point. In sizes larger than 10-point, the choice was standard: 12, 14, 18, 24, and 30-point, with 11, 13, 15, and 17 missing in Gothic 545, and 16-point available in only a few styles other than Gothic 545. As a result, a hierarchy of point sizes, consistent from map to map, is noticeable in the larger sizes.

As in the size characteristics of lettering on wax engravings, the style was dictated by the contemporary type styles available from the large American type founders who served the whole American printing industry, such as the American Type Founders Company. None of these styles was specifically designed for

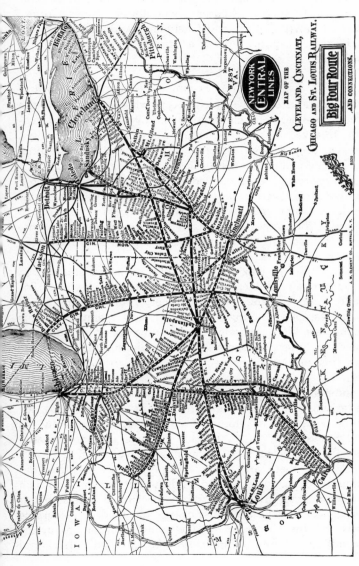

Fig. 5.10   Splitting of words in stamped lettering. See especially the Cleveland area. From *Atlas of Traffic Maps* (Chicago: LaSalle Extension University, 1930), p. 137

cartographic use, but a limited number of styles became popular for map work, such as Century Schoolbook, News Gothic, Stymie, Clearface Gothic, and Cheltenham. Occasionally, the use of swash capitals added a strangely stiff attempt at individuality.

Technical Peculiarities of Type

The wax engraver usually employed type trimmed as close to the letter as possible, thus eliminating the type shoulder. Occasionally this was not done, whereupon the type shoulder might be pressed deep enough into the wax to appear on the final printing plate and consequently in the map printed from it.

If the type were locked unevenly in the stamping stick, or if words were engraved in parts, uneven alignment of the individual letters sometimes resulted, giving a ragged appearance to the word. When type was "fingered-in" separately, this misalignment could also occur, but was usually less noticeable as the letters were further apart.

The versatility of wax engraving allowed letters stamped in with type to be readily embellished by hand with a stylus. For example, the engravers at Rand McNally routinely lengthened serifs on stamped lettering[14] (fig. 5.11), and examples of this procedure are also commonly found on wax-engraved maps produced by other publishers. It is probable that this technique was an attempt to relax the mechanical rigidity of type-produced names.

In summary, we find some characteristics of wax-engraved maps arising from the dominant use of foundry type for lettering. With the development of stick-up lettering in the United States about 1920, the use of type was perpetuated despite the decline of wax engraving relative to offset lithography. While there are many European atlases with hand lettering still being published today, the writer knows of no such major commercial atlas produced in the United States. But it should also be noted that the use of type has made considerable inroads into the European hand-lettering tradition, as in the recent editions of Philip's *University Atlas*, the *Times Atlas*, and others.

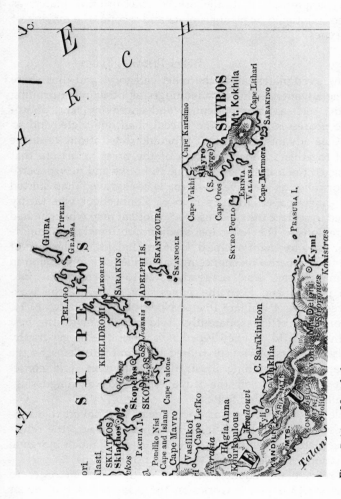

Fig. 5.11. Hand lettering of serifs. Compare the *K*s in "SKOPELOS" and "Skyros." From *Rand McNally & Co.'s Indexed Atlas of the World* (Chicago: Rand McNally and Co., 1892), p. 90

Chapter Five

PLATEMAKING

Certain characteristics of wax-engraved maps can be traced to the platemaking stage of the process. Two features were particularly important: first, the use of relief-printing plates, as opposed to intaglio or planographic; and second, the making of electrotypes.

Relief-Printing Plates

Wax-engraved plates fall in the relief category, and thus share the same advantages and disadvantages as other relief-printing methods, such as the woodcut, wood engraving, and photographic processes like the zinc linecut, halftone plate, and so forth. One advantage was that type and illustrations could be printed together in one operation of the press, a feature that encouraged the use of the technique in books and newspapers.

Historically, as we have seen, the wax-engraved map started out in a newspaper. Morse's map of Connecticut was tacitly intended to replace the type-inserted woodcut map which he had used previously. The newspaper medium continued to be important for the wax-engraved map. Even in England, at the turn of the century, where wax engraving had made little impression in other map forms, wax-engraved plates found their way into British newspapers. Even the *Times* saw fit to use the wax-engraved maps of William Phelps Northrup, of Buffalo, N.Y., and Lord Northcliffe apparently "told him he made the best maps in the world."[15] Encyclopedias, scientific texts, geography and history texts obviously shared the same problems as the newspaper of combining illustrations with text, and consequently capitalized on the advantages of wax engraving, especially in the United States.

While wax engraving was particularly well suited to newspapers or books, in which it is so commonly seen, the relief-printing plate also lent itself to the inclusion of much marginal information set in letterpress type around the maps.

The relief plate required pressure in the press in order to transfer ink to paper; it works on the principle of the rubber stamp. As a result, certain "impression characteristics" arise in all letterpress on relief-plate work, which contrast with those found in offset lithographic work, which is planographic, and hence does not require a heavy impression in order to print. Since the wax-engraved map is printed from a letterpress plate, it shares the impression characteristics of work produced by other relief processes, which can be conveniently identified as sharpness, high density, a tendency to show "ink squeeze," and indentations in the paper.[16]

Any comparison between early offset and letterpress work is almost certain to include words such as *sharpness, snap, sparkle, brilliance, luster,* and so forth, when referring to the letterpress impression.[17] The sharpness of wax-engraved maps arose from the bite of the plate in the paper, producing sharp, well-defined edges to the lines, lettering, and tints. It is particularly noticeable when the work was printed on glossy or coated paper.

Closely associated with the characteristic of sharpness, the opacity of letterpress inks has also drawn comment, such as that of the "old-times pressman" quoted in an article comparing offset and letterpress work: "If you want to put guts into printing, you've got to use good ink and put it on paper by the letterpress process."[18] A more elegant view was expressed by an experienced engraver in one of the interviews made for this study:

> I think it took them years till they got offset down to the point where it was acceptable to the publisher. They experimented around with it, but the public just couldn't accept the grayness of one signature compared with another in a book. About 20 to 25 years ago, we experimented with some offset for Ginn and Company. We gave them some artwork, and they prepared a pre-publication pamphlet on the book they were to publish, but they just couldn't get the pages to look the same. They were disgusted with it. So we went through the whole series using wax-engravings. I think they had a hard time accepting offset, because of the quality primarily. You really got sharp lines with wax-engraving. You can't beat it, of course. It's impossible to beat the sharpness and blackness of wax-engraving.[19]

When solid colors were used on wax-engraved maps, the characteristic opacity of letterpress inks was particularly apparent. It was not possible to achieve a transparent effect, such as that frequently attained by lithography, without the use of extremely narrow ruling. Even if a light value ink were used, it still possessed an opacity which covered existing detail underneath.

Most letterpress impressions other than those of the highest quality exhibit at least some "ink squeeze."[20] This occurs when the printing surface under pressure forces excess ink to escape sideways from under the plate, leaving a residue of ink on the outer edges of the impression. When the printing surface extends over a large area without a break, as in a solid color, the squeezed ink has no opportunity to escape, and consequently dries in a mottled pattern, a striking characteristic of a solidly inked area on a map. Ink squeeze is more prevalent on any nonabsorbent paper, such as glossy or coated paper, where the phenomenon can be caused by the slightest overinking in the press.

An obvious characteristic of letterpress printing is indentations in the surface of the map caused by the impression. These may often be discovered by feeling the back of the map, or by holding it up so that the light reflects off the ridges. On a multicolor map, with registry marks on the corners, such indentations may be easily felt (provided that the colors are in good register), as they arise from repeated impression from different color plates on the same spot. Where it is difficult to identify the use of a relief-printing process, the author has found this method to be particularly satisfactory.

When Sidney Edwards Morse was introducing his invention, cerography, to the readers of the *New York Observer*, he went into some detail to describe the advantages and disadvantages of the three major printing processes of the day: lithography, copperplate engraving, and wood engraving. In comparing cerography with these methods, he was careful to refer to the speed of printing with relief plates in a power press: "The printing is executed with the common printing press, and of course as rapidly as woodcuts or letter-press printing; that is, at the rate of 6,000 square feet in ten hours, for beautiful work under the hand-press, or 60,000 under the single Napier."[21] Morse estimated this speed of printing to be sixty times faster than lithographic work, and a hundred times faster than copperplate printing, which of course were limited to the hand press until the rise of rotogravure and offset lithography at the end of the nineteenth century. This dramatic increase in the speed of presswork naturally resulted in a corresponding decrease in the cost per printed sheet. This is reflected in Morse's ability to give away a number of atlas supplements to the *New York Observer*. He had estimated that the use of copperplates in this project would have cost $125,000 for 17,000 copies, which amounts to about $7.35 unit manufacturing cost, high indeed at that time.[22]

The ability of wax-engraved plates to be used in the "common" printing press was only one factor in providing cheaper maps for the public. Cheap paper was another. Generally speaking, the importance of paper in the development of cartography has been largely ignored, except in the use of watermarks to date them as historical documents. It is surprising that the most basic physical material of the map should not have been treated. It is sometimes forgotten that, like the wood engravings of Bewick,[23] high quality lithography would not have been possible without the introduction of smooth "wove" paper in the mid–eighteenth century, and that the development of commercial book and map production in the nineteenth century would have been considerably curtailed but for the Fourdrinier papermaking machines developed at the close of the eighteenth century.[24] Similarly, the use during the nineteenth century of materials other than expensive rags in the manufacture of paper was a prerequisite for cheaper maps.

Significantly, the first commercial use of wood pulp in paper manufacture and the invention of wax engraving were contemporary events. In 1839 Charles Fenerty is said to have experimented with ground wood as a papermaking substance, and in 1841 he claimed to have produced a durable, white, firm paper from spruce. In 1840 Friedrich Gottlob Keller was granted a German patent for a wood-grinding machine, which was bought by Heinrich Voelter in 1846 and developed for the mass production of wood pulp.[25]

The wax-engraved printing plate was able to take advantage of the development of cheaper paper for the reason that it was an extremely deep, strong form of relief-printing surface. The building up of the spaces between the engraved lines produced a deep "hold," and the flaming procedure (a brushing of the wax surface with a gas flame) flowed the wax right up to the engraved lines, strengthening the resulting electrotype. In addition, the V-shaped engraving tool produced strong lines in the electrotype. The deep, sharp character of the wax-engraved printing plate produced an acceptable impression on the softest and coarsest cheap paper, such as newsprint, a claim by no means all printing processes could make, even if they were relief techniques. For example, photographic linecuts were characteristically shallower than wax-engraved plates, and the edges of their printing surfaces were more rounded, so that ink could be picked up on the shoulders and cause an unsightly ragged edge to the image when printed under pressure on soft paper. Such plates produced more satisfactory work on the harder coated papers, which were more expensive and, one might add, considerably heavier in large sizes.

The use of wax-engraved plates on power presses fed with cheaper paper released a large amount of money to be channeled to other destinations. While several of these destinations were undoubtedly the pockets of printers and publishers, it can also be stated that the map buyer began to receive more for his retail dollar. One such benefit appears to have been the frequent addition of color to wax-engraved maps.

Color printing has always been notoriously expensive. Much of this cost arises from the need for a separate plate for each color, each requiring a separate press run, with the result that the cost of printing four colors is approximately four times that of printing one color, plus the additional expenses of precisely registering the plates and making other adjustments. Until the nineteenth century, the printing of maps in one color was expensive enough; the late appearance of maps printed in multiple colors is therefore hardly surprising.

Wax engraving was both able to produce color plates and take advantage of cost reduction by printing them on power presses. For this reason, color is commonly seen on even the earliest wax-engraved maps, such as those of Sidney Edwards Morse in the *Cerographic Atlas of the United States*, the *Cerographic*

Bible Atlas, and so forth. Later patents, such as that of Charles H. Waite, set out improvements in the use of color on wax-engraved maps by the combination of two or more primary-color–ruled tint plates to produce other colors.[26]

The relief plate is particularly susceptible to damage, especially when it is being constantly taken out of storage for reprinting purposes, as was commonly the case with maps. The most vulnerable parts of the plate were the sharp edges of border lines, which could be easily bent, rounded, or otherwise distorted. Such damage did much to transform sharp and regular wax-engraved lines of a new plate into ragged and unsightly work.

Electrotyping

Many characteristics of the wax-engraved map arose from the use of electrotyping as the basic platemaking procedure. For the purposes of this book, electrotyping is considered to include not only the process of electrodepositing metal on the mold, but also the subsequent significant steps of *backing up* and *electrotype finishing*. The electrotyping stage is considered to have contributed characteristics to the wax-engraved map in matters of registration, size of plate, length of run, and ability to make corrections.

The most casual observer looking at a wax-engraved map cannot fail to notice that the colors frequently do not "fit," or are out of register, a fault usually blamed on the printer. But wax-engraved plates were often impossible to register precisely, because of differential shrinkage of each plate during the electrotyping procedure. Such shrinkage rates could not be precisely controlled by the electrotyper. The shrinkage occurred when the electrotype shell was backed with molten type metal. On cooling, the whole plate would shrink about one thousandth of an inch for each inch of plate. This amount was allowed for by one engraver interviewed by the writer.[27] The experience of Rand McNally is also interesting in this regard:

> Shrinkage was one of the unfavorable characteristics of wax engraving and was the prime reason for the poor fit or register of maps made by this method. The amount of

shrinkage would run from 1/32 of an inch on the smaller maps up to 1/8 of an inch on the big, solid color sheets and would often vary from plate to plate. On maps requiring more than one plate for the base map, the "joins" were extended 3/16 of an inch to allow for shrinkage. To compensate for the variation in shrinkage in the numerous plates necessary to produce the map—linework, colors, patterns—it was often necessary to cut the copper engraving plate into pieces, usually along country, state or county boundary lines as these cuts would be less noticeable.[28]

While Sidney Edwards Morse stated that "we know of no limit to the size of cerographic plates," he realized that there was an upper limit set by the size of the bed on the press to be used and consequently implied quite correctly that, should one wish to print anything bigger than the bed of the largest press available to him, he should print it in pieces.[29] Similarly, although the common electrotyping tank could hardly handle a plate larger than about one or two feet square, this problem could be readily solved by making up the plate out of a number of small electrotypes nailed or screwed securely to a wooden base. This method did, however, give rise to characteristic breaks in the line work on a wax-engraving map, as described by one engraver: "It is surprising, today we are so critical about register and any little opening in lines. In those days (of wax-engraving) you had the width of a saw blade opening between your lines. That was it, that was standard. They were never fixed."[30]

The extremely good wearing qualities of an electrotype contributed to the cheapness of presswork which was discussed above. The Research Committee of the Association of Electrotypers and Stereotypers reported in 1940 that a copper electrotype could be expected to yield up to 150,000 good impressions, while the nickel electrotype would yield up to 250,000.[31] In addition, chromium plates lasted from two to five times longer than the copper or nickel plates. In the early days of wax engraving, Morse claimed that a million good copies could be struck from one of his cerographic plates in contrast to the one or two thousand possible from a lithographic stone, while the copperplate engraving had to be retouched after 10,000 impressions.[32] Although we may regard Morse's claims to be somewhat exaggerated, the electrotype certainly had advantages of durability in the press. In addition, should more copies be required, stereotypes could be made from the original electrotypes at an extremely low cost compared to transferring a drawing to a new stone, or electrotyping a new copper intaglio replacement plate. The immediate effect of these advantages was to provide the public with cheaper maps, often embellished with printed color.

It has been pointed out in the literature that wax-engraved plates were well adapted to the making of corrections and revisions, as in the account by Fitchet:

> The surge in new settlements certainly sold thousands of maps, but as more and more new towns were developed, a method had to be devised for making rapid corrections, inexpensively, to keep new editions up to date, and to make new maps as the need would arise. The face of our land was changing daily.
>
> A technological change solved this problem when Rand McNally abandoned the expensive and inadequate methods of engraving and re-engraving on copper, steel and stone. The wax-engraving process permitted the insertion of correction patches in the printing plates. Maps could be corrected very quickly and reprinted frequently at but a small fraction of former costs. This, of course, gave great impetus to map sales.[33]

While a copper engraver could make small changes in an intaglio plate by hammering the old detail flat, beating up the surface of the area level with the remainder of the plate, and reengraving new detail, there was a limit to both the number of times this could be done and the size of the area that could be corrected. Large corrections were also difficult for the lithographer. In wax engraving, however, a whole part of the electrotype could be sawn out of the original plate and repatched with a new wax engraving of the corrected detail.[34] These patches usually fitted exactly, and were very difficult to detect. This stage of the process was part of the work of the electrotype finishers, an extremely skilled class of craftsmen, who could hammer and tool a damaged or corrected plate until it printed like new.[35]

123

6 Postscript

Compared to other commercial map-printing processes, wax engraving found much wider use in the United States than in any other country. As we have seen, the technique appears to have been largely responsible for producing a typical American atlas cartography, especially during the main period of the technique from 1870 to 1930. Moreover, the technique seems to have left some vestiges of its existence. While there was a significant change in style as publishers went from wax engraving to offset lithography, wax engraving appears to have left some imprint on the style of American commercial atlases being produced today, despite the use of a different process. Thus, Stanley Larson, for some time a wax engraver and still in the graphic arts business as a professional illustrator, had this to say: "I think it [wax engraving] definitely was a factor in setting style, because, I'll be honest with you, we use the same procedure now. The black base is the same. We use foundry type, only it's printed on transparent material. The black base is the same. So little difference."[1] The map extract in figure 6.1, produced by the American Map Company (which never used wax engraving), illustrates the striking similarity between it and the typical examples of wax engraving which are illustrated throughout this book. Thus, Erwin Raisz's observation made in 1948 concerning the characteristics of American commercial cartography still appears to be valid now: "Wax engraving became the universal American method of map reproduction, and the over-lettered and mechanical-looking maps still are characteristic of American commercial cartography."[2]

One reason for the apparent continuation of a cartographic style reminiscent of wax engraving is that the original wax-engraved plates were often transferred to film to make photo-offset plates, or photoengraved relief line plates. The latter is still the practice of the George F. Cram Company. While the author was touring the Indianapolis plant of that firm late in 1968, he was shown a large wax-engraved base plate of the United States that was used to make proofs which were then photographed and corrected on the film. With such a procedure, at least some of the characteristics of the original engraving will obviously continue.

Nevertheless, there is also evidence for vestiges of the style characteristic of wax-engraved maps in modern productions from new compilations in which wax engraving plays no direct role. It is as if there is a traditional image in the mind of the contemporary mapmaker, gleaned from the days of wax engraving, that somehow pervades the modern American atlas map. Whatever the reasons, the majority of American commercial map-printing establishments continue to produce a map that differs in style from the European model. The difference is openly admitted, and there is a continued respect for the products of most European cartographers. An American map company, in their 1969 catalog, carried a line of Bartholomew wall maps made in Britain which they described as "a truly sophisticated school map line for the discerning teacher who appreciates classical qualities in map making," which tacitly differentiates them from the company's own products. The question naturally arises as to whether this particular company is also capable or desirous of producing a "map line for the discerning teacher," or whether it wishes to continue in its implied policy of providing maps for the less discerning.

It is interesting to speculate on the effect of bombarding generations of American map users, in school, home, and office, with maps and atlases of the particular style nurtured by wax engraving. Teachers and school boards, representing a large part

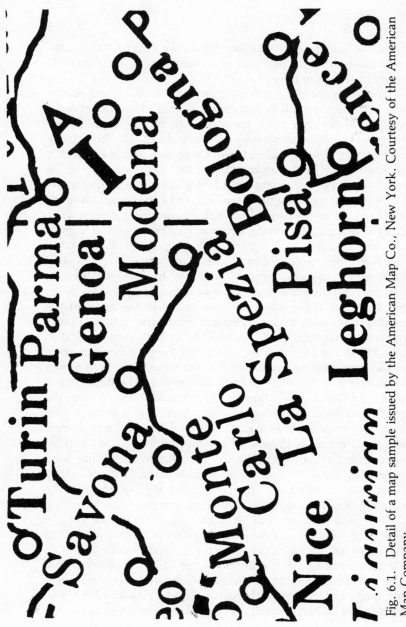

Fig. 6.1. Detail of a map sample issued by the American Map Co., New York. Courtesy of the American Map Company

of the clientele, certainly became satisfied with them, perhaps because they had neither the training nor the imagination to demand something better. In addition, any charge leveled at the map and atlas firms for continuing to produce mediocre material could be answered in economic terms with reference to the demand. Thus, since both supplier and client were satisfied with the status quo, there existed no opportunity for fresh and intelligent ideas to be fed into the system. Neither was there even a potential source of fresh and intelligent ideas, not only because they were not encouraged, but also because there were few educational institutions in the United States which could provide the basis of a cartographic education on which such ideas could be formulated.

A famous example of the blind acceptance of mediocre material on the part of teachers and school boards is found in the enthusiastic support for the use of the Mercator projection for general reference world maps in school atlases. This projection, with its enormous distortion of area in the higher latitudes, has been responsible for many misconceptions on the part of many people. Could it also be that the typical wax-engraved map, with its stiffness in line work and lettering, its dull, flat appearance, and its attempts to represent terrain with feeble hachures, contributed to the stiffness and dullness of the geography text it sought to illustrate?

Appendix

Contracts relating to the early history of wax engraving, between Sidney Edwards Morse, Samuel Breese, Alfred Munson, and Harper and Brothers (17 April 1844, 15 March 1845 [2], 17 March 1845), and a memorandum from Alfred Munson to Sidney Edwards Morse (20 November 1846). These documents are in the archives of Harper and Row, 49 East 33rd Street, New York 10016.

[17 April 1844]

This Indenture, made the seventeenth day of April, A.D. Eighteen hundred & forty four, between Harper & Brothers, of the first part, & Sidney E. Morse, & Samuel Breese, of the second part, all of the State, County & City of New-York, Witnesseth,

That said Sidney E. Morse and Samuel Breese, the party of the second part, having prepared the copy of a School Geography fitted for the use of pupils in the Common Schools throughout the United States, and illustrated by maps from Cerographic plates, engraved & constructed by them, it is hereby agreed, that said Harpers & Brothers shall have the exclusive right of selling said geography for the term of the Copyright, on the terms following, viz.

1. Said Harpers and Brothers shall pay said Sidney E. Morse, within thirty days after publication Fifteen hundred dollars, being one half of the estimated value of said plates.

2. Said Morse & Breese shall provide coloring plates, & said Harpers shall do the composition and provide the necessary wood-engravings & stereotypes of said Geography at cash-prices on joint account.

3. The book shall consist of about 80 quarto pages of the medium size done up in the usual School Atlas form, and the retail price shall be fifty cents a copy.

4. The book shall be manufactured, according to a sample to be

129

mutually agreed upon, and by the said Harpers, at the market-price, for paper, press-work & binding, unless said Morse & Breese, or either of them, are willing to do it at the same rate. And before printing each edition, said Morse & Breese (or either of them) are to have the privilege of saying whether they will manufacture that edition or not.

5. Said Harpers agree to keep the market in all parts of the United States fully supplied with said Geography and always to sell them by wholesale for One thousand copies & upwards on the usual credit, at not less than thirty per cent. discount from the retail price.

6. Said Harpers agree to pay said Morse & Breese for the copyright of each edition of said book, one half the difference (5 per cent. less, guarantee) between the manufacturing cost and the wholesale price (viz. 35 cents a copy) the payment for each edition to be made in a note at six months from the time that the first copy of the edition is ready for sale. If said Morse & Breese prefer it, they may receive in lieu of the note at six months, books at the manufacturing cost, at the time of publication with no deduction for guarantee.

7. Said Morse & Breese engage to read the proof-sheets, revise, correct and rewrite said geography whenever necessary and to use their best talent in devising & inventing such improvements as may tend to preserve & extend the reputation of the work. And they further agree not to engage or be interested in the publication of any other School Geography which will interfere with this in the market; it being understood that a geography which sells to the trade at wholesale for .15 cents, or less, or for 75 cents & upwards, will not be regarded as an interference. Should said parties of the second part be entirely satisfied with the results of said Geography, in such case Harpers are to have the refusal for publishing said other geographies.

8. Said Harpers engage that they will not be concerned in the publication, or sale of any other School Geography or Atlas that will interfere with this work, and that they will do all in their power to promote its sale by causing the book to be advertised extensively in all parts of the United States and taking vigorous measures to have the book presented to school teachers & trustees throughout the land. One thousand copies to be given away, are to be charged to joint account.

9. And when new Cereographic [sic] plates shall be wanted, said Morse & Breese shall cause them to be engraved & constructed at a profit of not more than thirty three per cent. on the cost to them, to be charged to the profits of the work.

In presence of
G. L. Demarest.

Sidney E. Morse
Samuel Breese
Harper & Brothers

Appendix

Articles of agreement between *J. Harper & Brothers* of the first part, *Sidney E. Morse* of the second part, and *Henry A. Munson* of the third part, all of the City of New-York.

1. Said Munson to superintend a Cerographic Establishment for said Morse, and said Morse & Harpers for three years, at the rate of one thousand dollars a year, payable monthly.

2. The duties of said Munson will include, First, aiding as far as in his power, said Morse in his experiments for the perfection of Cerography. Secondly, training such persons as may be selected by said Morse to the business of engraving & finishing cerographic plates: thirdly the general superintendence of the business: and fourthly, the execution of the more delicate processes of the Art, so far as can be done consistently with the duties already named.

3. Said Munson to keep in a Book to be provided for the purpose, an accurate account of the time of all persons employed in the business: carefully separating the time spent in executing plates for use by said Harpers, or by said Morse & Harpers, from the time spent in making plates for others than the said Harpers. Ten hours to constitute a day.

4. All time spent in making Experiments for the improvement of the Art, or in making plates in which the said Harpers have no interest, to be paid for wholly by said Morse: but all time spent in making plates, or in executing work in which the Harpers are interested to be paid for as may be agreed upon between said Harpers & said Morse.

5. Said Munson not to disclose any of the secrets or processes of Cerography, or any of the Experiments above referred to: and not to use or train others to use any of the processes connected with Cerography, except with the written permission of said Morse separately, or of said Morse and Harpers, jointly.

6. In case Cerography should be superceded [sic] by some new invention, said Munson agrees to occupy himself with other and kindred employments, under the direction of said Morse and Harpers, at the same rate of wages: or to release them from the obligations of this contract, on the payment of Two Hundred dollars at their option.

7. Mr. Munson to keep an account of time devoted to Mr. Morse's experiments.

New York, Mch. 15, 1845

Henry A. Munson
Sidney E. Morse
Harper & Brothers

131

Harper & Brothers

[15 March 1845]

Articles of Agreement between *Sidney E. Morse* of the first part, & *J. Harper & Brothers* of the second part.

1st. Said Morse to prepare as fast as he conveniently can, the copy of Maps for a Primary Geography, & of State Maps of any of the States of the American Union, and to engrave them by Cerography. Said Harpers engage to publish them, as fast as engraved, on the principle of half-expense, & half profits. Said Morse to be paid by said Harpers for their part of the engraving & plates one half of the actual cost of labor, rent, fuel, light, & materials, with the addition of thirty-three & one third per cent. as his engraving profit.

2. The Harpers having purchased Mr. Breese's interest in the School Geography are entitled under the School Geography contract to three fourths of the profits of the work, & will pay hereafter to the said Morse, three fourths of the actual cost of engraving and manufacturing new plates, & repairing old plates, with an addition of 33 1/3 percent. on the cost of labor, rent, fuel, light and materials.

3. The Harpers having purchased Mr. Breese's interest in the North American Atlas, are owners of one half of that work, & the work being in an unfinished state, it is mutually understood and agreed that it shall be finished forthwith, and extended to a General Atlas, on the terms following: viz:

First. The duties of authorship shall be assumed and discharged wholly by said Morse.

Secondly. The present plates shall be finished and new plates shall be engraved by said Morse for said Morse & Harpers, at cost: with the addition of 33 1/3 percent. as in the case of the state maps.

Thirdly: As soon as a sufficient number of plates are finished, the work shall be printed on paper, and in a style according to a sample to be mutually agreed upon, & by said Harpers, at the market price for paper, press work, & binding, unless said Morse is willing to do it at the same rate; and before printing each edition, said Morse is to have the privilege of saying whether he will manufacture that edition or not.

Fourthly. Said Harpers shall keep the market in all parts of the United States fully supplied with said atlas, and always sell it by the hundred copies at 40 per cent. discount from the retail price.

Appendix

Fifthly.   Said Harpers agree to pay said Morse for the copyright of each edition of said atlas, (or of each number of said atlas, if published in numbers) between the market price of manufacturing and wholesale price of i.e., price by the hundred copies. Accounts for copyright to be settled on the 1st. February & 1st. August in each year by note at four months.

4th. The principles of the preceding (or 3rd.) Section to apply to the Primary Geography, and to the state maps.

5th. Said Morse assumes all the duties of authorship now belonging to Morse & Breese in the School Geography contract: and said Harpers agree that if any damage shall accrue to their interest in said Geography, or to their interest in any other Cerographic work, from inability to command the aid necessary to the execution of Cerographic plates, they will release said Morse, and they do hereby release him from all claim for such damages, he using all reasonable diligence to prevent it.

Given under our hands & seals this fifteenth Day of March, A.D., 1845 in the City of New York.

[The words "for copyright" in the 4th. line from bottom of 2d. page inserted before signing]

Harper & Brothers,
Sidney E. Morse

[17 March 1845]

Articles of agreement between *Sidney E. Morse* of the first part; *Samuel Breese* of the second part; and *James Harper, John Harper, Joseph Harper & Fletcher Harper*, doing business under the name of *Harper & Brothers*, of the third part, all of the City of New York.

Whereas said Morse, while engaged in preparing maps for the subscribers to the *New-York Observer*, between the years 1834 & 1839, invented a new mode of engraving & obtaining in metallic relief plates fitted for printing, under the common printing-press—which new mode or Art is called *Cerography*; and whereas near the close of the year 1839, he disclosed the secrets of this art to said Breese, & has since been engaged in connection with said Breese in getting up Cerographic plates for *Morse's School Geography*, and for a *North American Atlas*, published, or to be published by said Harpers; Now, therefore, it is agreed by & between said Morse & Breese, & by & between all the parties to this contract, as follows:

1. Cerography, & its apparatus, & all other joint property of said Morse & Breese, except Morse's School Geography, & the plates of the

133

North American Atlas are to be, & they are hereby made the sole and exclusive property of said Morse, & for his sole & exclusive use.

2. Said Breese shall sell by another instrument, bearing even date with this, to said Harper & Brothers all his right, & interest in Morse's School Geography, for the sum of Five Thousand dollars to be paid in Cash on the Execution of the instrument.

3. Said Morse shall have the right to use the plates of the North American Atlas for the fulfillment of existing pledges to the subscribers to the New-York Observer, without charge.

4. Said Breese in another instrument, bearing even date with this, after reserving to said Morse the right to use the plates of the North American Atlas for the fulfillment of existing pledges to the subscribers of the New York Observer without charge, shall sell his interest (being one half) in these plates to said Harpers for the sum of Three Thousand dollars, to be paid at the end of three years after the plates (including the coloring plates) are finished and delivered ready for the press, with interest annually at six per cent during the three years. And it is mutually understood and agreed that said Morse & Harper are to be at the expense of finishing said plates, and are to grant every reasonable facility in their power for the purpose; but said Breese is to correct the proofs of the unfinished plates without charge, and to be at all risks, so far as relates to his half of these plates, until they are finished and delivered ready for the press.

5. Said Breese shall prepare for the engravers in the course of the year 1845, if his health will permit it, in a style equal to that of "Mitchell's Map of the U.S." (and with the addition of the county lines, and names of the counties, so far as they can be ascertained) the drawing or copy of a map of the United States, four feet long by three feet broad, including the border; and within one month after the delivery of the finished copy of the map by said Breese, said Morse and Harpers shall commence engraving the same by the Cerographic Art, and shall proceed with all reasonable dispatch until the plates are wholly and neatly finished. And they shall then print the map in a workmanlike manner, on good paper, & shall use their best endeavors to sell it, and shall allow said Breese as a full compensation for his labor, ten cents a copy on each of the first Twenty Thousand copies sold. And said Morse and Harpers agree not to publish, within three years from this date, any Cerographic or Anastatic [i.e., lithographic transfer] Map of the United States that will interfere with the Map so prepared by the said Breese, until he shall have received his Two Thousand dollars under this arrangement, it being understood that a

map varying more than six inches each way in size, will not be regarded as an interference.

The phrase "reasonable dispatch" in the first paragraph of this article is understood to refer especially 1st. to the difficulties of Cerography, as a new and secret Art, practised in all in [sic] parts by only one individual, whose health is not firm, & whose services Messrs. Morse & Harpers may not, therefore, be able to command; and 2nd. to the exigencies of the School Geography, of the North American Atlas, and of the Maps necessary to fulfill existing pledges to the subscribers to the New York Observer—all of which are to take precedence of the Map of the United States; it being regarded as reasonable that Messrs. Morse and Harpers should not be required to neglect these for the sake of any new work. If, however, at the end of three months from the delivery of the finished copy of the map by said Breese, or at the end of any subsequent period until the work shall be wholly completed, said Morse & Harpers shall not have employed an amount of labor upon the work equal to the constant labor of one hand during the working hours of each working day, the said Breese shall then have the right to engrave the map on Copper or Steel, and to sell the same wholly on his own account.

6. Although it was not intended, and is not intended, in the composition of any of the maps here referred to, to take any portion of the same from any maps or works secured by Copyright, so as to infringe said Copyright, yet as it may happen that claims for infringement may be made, said Breese hereby agrees that if in consequence of any violation of copyright in preparing said School Geography, North American Atlas, or Map of the United States, said Morse & Harpers, or either of them shall suffer loss or damage, the said Breese will make them good to the extent of one half of the loss or damage suffered from the violation of copyright in preparing the School Geography or North American Atlas, & the full extent of the whole loss or damage which they or either of them may sustain from any violation of copyright in preparing the Map of the United States.

And whereas said Breese is unwilling at present to guaranty said Morse & Harpers against loss from the violation of copyright in preparing the following seven maps or plates of the North American Atlas viz. 1. New Hampshire and Vermont: 2. New York, & vicinity: 3. Georgia: 4. Alabama: 5. Mississippi: 6. Louisiana: and 7. Ohio: it is hereby mutually understood and agreed, anything in this article to the contrary notwithstanding, that the said Breese may have the privilege of declining altogether the guaranty of any of these seven plates: he giving written notice to said Morse & Harpers of his determination to do so at any time before the plates of the North American Atlas are finished and delivered, and deducting from the allowance to him for said Atlas, Eighty Dollars for each of the plates which he shall decline to guaranty.

7. Said Breese hereby, and in consideration of the stipulations of this agreement, binds himself that has not reveal [sic], and will not reveal, directly or indirectly, by hint, innuendo, or in any other way any of the secrets or processes of Cerography, & that he will not use the same, & that he will not engage, or be interested in the preparation or publication of any maps or prints from anastatic plates, or from plates in metallic relief (except from metallic stereotypes of wood cut engravings and letter type, or anastatic transfers from letter type) under penalty of forfeiture to said Harpers of all his interest in this contract, & of forfeiture to said Morse of Twenty Thousand dollars, liquidated damages.

8. Said Breese here declares it to be his intention, after preparing the Map of the United States herein referred to, to retire from the Map & Engraving business mentioned in Article 7, & to leave said Morse and Harpers in the quiet & peaceable possession of the whole field. And said Breese hereby agrees that he will not visit the rooms in which experiments are performed, or any Cerographic process or business is carried on, for said Morse, or for said Morse & Harpers, without their written permission; and he further agrees that he will not converse with anyone who may have been in said Morse's or said Harper's employment, in relation to the experiments of said Morse, or in relation to the business connected with Cerography, of said Morse or said Harpers, but will truly and faithfully avoid all conversation and correspondence with everyone, and all conduct, which conversation, correspondence, or conduct may tend to prevent the said Morse, or the said Harpers from reaping the full and exclusive benefit and advantage of Cerography (or of any kindred mode of obtaining plates in metallic relief from engravings), in all its application to maps, books & prints of every description. [The contract has no article 9.—Author]

10. Said Morse shall supply said Breese's correspondents with all the maps and atlases promised for their aid in furnishing map materials, provided the expense does not exceed One Hundred and Fifty Dollars.

11. After the signing of this instrument, said Morse shall give up all claims against said Breese, both on Cerographic & private account & mutual receipts shall be exchanged in full of all demands, except such as may arise from this contract & the other contracts of contemporaneous date herein referred to.

Appendix

12. No single maps from the plate of New Jersey, or from the plate of Tennessee & Kentucky, to be published by Morse & Harpers, except at their own risk, and said Morse not to be required to furnish said Breese's correspondents with single maps from these plates, anything in the tenth article to the contrary notwithstanding.

New-York. March 17, 1845

Samuel Breese
Sidney E. Morse
Harper & Brothers

[20 November 1846]
[From a memorandum from Henry A. Munson to Sidney E. Morse]

N.B. The first 16 maps for the North American atlas were delivered June 25th, 1845. 16 more in May 1846. The last four early in July 1846.

In the fall of 1845 the North American Atlas was laid aside at request of Messrs. Harpers and an unsuccessful attempt made to get out the plates for the School Geo. with the understanding that this delay should not cause any loss to Mr. Breese.

As the first 16 maps were finished in about 4 months, the other 20 could have been completed in 5 months more, making nine months from the 14th of March 1845. Provided no other work interfered and the operations of Cerography had been successful—

N.Y. Nov 20th 1846

Henry A. Munson

1. It is less than two hours work to prepare a plate of the size of N. A. atlas for the engravers.
2. The colouring plates for N. A. Atlas were engraved by persons without any experience in Cerographic engraving. The average time spent on each was something over a day. Some few were spoiled from inexperience.
3. The time required by Mr. Munson for each colouring plate averages two days of about six hours labour each day. No plate was lost by him either in casting or otherwise.
4. Each colouring plate requires 1/2 a day of a common hand.*
5. It required of Mr. Munson's time to complete the maps as follows:

N. America	4 days
Maine	4
Pennsylvania	3
Kentucky & Tennessee	3
Missouri	2
Georgia	1/2
Arkansas	2

137

Louisiana	2
Texas	3
Mexico	3
West Indies	1
	27 1/2

*[A "colouring plate" was a wax plate in which fine tints were ruled to shade down color. One plate for each color was necessary.—Author]

Colouring Plates

West Indies	2	Missouri	2	New York City	0
Mexico	2	Kentucky & T	2	do & vicinity	0
Texas	2	Virginia	2	New York	2
Louisiana	2	Illinois	2	Connecticut	2
Florida	2	Indiana	2	Vermont &	
Indian Territory	0	Ohio	2	New Hampshire	2
Arkansas	2	Maryland	2	Maine	2
Mississippi	2	Wisconsin	2	Nova Scotia	2
Alabama	2	Iowa & Wis	2	Canada West	2
Georgia	2	Michigan	2	Canada East	2
S. Carolina	2	N. Jersey	2	N. America	2
N. Carolina	2	Pennsylvania	2		16
	22		24		24
					22
					62

Add 27 1/2 days
on Maps   27 1/2
              89 1/2

Notes

Chapter One

1. Arthur H. Robinson, "Mapmaking and Map Printing: The Evolution of a Working Relationship," in David Woodward, ed., *Five Centuries of Map Printing* (Chicago: University of Chicago Press, 1975).

2. "A Map of New-England ...," in William Hubbard, *A Narrative of the Troubles with the Indians in New-England* ... (Boston: John Foster, 1677).

3. This discussion and the paragraphs that follow are based on Woodward, *Five Centuries*, to which the reader is referred for further general information on the history of map printing techniques.

Chapter Two

1. "Electrolysis," in *Encyclopaedia Britannica*, 11th ed. (1910).

2. F. J. Wilson, *Stereotyping and Electrotyping*, 6th ed. (London, 1898), p. 98.

3. The early history of electrotyping is reviewed in some detail in C. S. Smith, "Reflections on Technology and the Decorative Arts in the Nineteenth Century," in Ian M. Quimby and P. A. Earl, eds., *Technological Innovation and the Decorative Arts*, Winterthur Conference Report, 1973 (Charlottesville, Va.: University Press of Virginia, 1974), pp. 1–64.

4. Thomas Spencer, *An Account of Some Experiments Made for the Purpose of Ascertaining How Far Voltaic Electricity May Be Usefully Applied to the Purpose of Working in Metal* (Liverpool: Mitchell, Heaton and Mitchell, 1839), pp. 13–14.

5. Wilson, *Stereotyping*, p. 98.

6. Wilson, *Stereotyping*, p. 104.

7. William Tudor Mabley, "Certain Improvements to Be Used in Printing, Embossing, or Impressing," *Patent Specification no. 8743*, London, British Patent Office, 17 December 1840.

8. Ralph H. Brown, "The American Geographies of Jedidiah Morse," Association of American Geographers, *Annals* 31 (1941): 145–217.

9. *New York Observer*, 29 June 1839, p. 102.

10. *New York Observer*, 2 January 1841, p. 2.

11. Richard Cary Morse to Sidney Edwards Morse, 22 October 1845, Morse Family Papers, Yale University Library, New Haven, Conn.

12. Ibid.

13. Eugene Exman, *The Brothers Harper* (New York: Harper and Row, 1965), p. 235.

14. Sidney Edwards Morse to Richard Cary Morse, 7 August 1845, Morse Papers, Yale University Library, New Haven, Conn.

15. Edward Lind Morse, *Samuel F. B. Morse* (Boston, 1914).

16. Ernst Fischer, "Zweihundert Jahre Naturselbstdruck," *Gutenberg Jahrbuch* (Mainz: Gutenberg-gesellschaft, 1933), pp. 186–213; and Roderick Cave and Geoffrey Wakeman, *Typographia Naturalis* (Wymondham, England: Brewhouse Press, 1967).

17. Eugene Exman, *Harper*, p. 184.

18. John Alfred Nietz, *Old Textbooks* (Pittsburgh: University of Pittsburgh Press, 1961), p. 224, quoting John R. Sahli, "An Analysis of Early American Geography Textbooks from 1784 to 1840" (Ph.D diss., University of Pittsburgh, 1941); and Ned Culler, "The Development of American Geography Textbooks from 1840 to 1890" (Ed.D. diss., University of Pittsburgh, 1945).

19. For example, Morse's "Cerographic Map of Wisconsin, by Charles W. Morse" (Chicago: Rufus Blanchard, 1855).

20. George F. Cram Co., Typewritten notes concerning the history of the company (no date).

21. George H. Benedict, "Map Engraving," *Printing Art* 19 (1912): 205.

22. William Haas, "The Wax Process," manuscript (ca. 1920), pp. 4–5.

23. J. Paul Goode, "The Map as a Record of Progress in Geography," Association of American Geographers, *Annals* 17 (1927): 12.

24. George Sherman, "Outgrowths of Letterpress, no. 8: Modern Mapmaking," *Inland Printer* 55 (1915): 321.

25. "Wax Process of Engraving," *Inland Printer* 68 (1922): 650.

26. Andrew McNally III, *The World of Rand McNally* (New York: Newcomen Society of North America, 1956).

27. Rand McNally and Co., "First Maps," manuscript, p. 3.

28. Rand McNally and Co., *Railway Guide* (December 1872), p. xxii.

29. Rand McNally and Co., *Railway Guide* (April 1873), p. xv.

30. Charles B. Frey, Jr., to David Woodward, 17 October 1968.

31. "Some Railway Map-Making," *Inland Printer* 15 (1895): 500.

32. *Bulletin of the American Geographical Society* (1900), pp. 168–69.

33. Rand McNally and Co., Manuscript relating to the history of the company (no date).

34. "Scales and Stars," *Inland Printer* 99 (1937): 37–40.

35. "A Loss to Publishing," *Publishers Weekly* 116 (1929): 2926.

36. Rand McNally and Co., *Pioneer Atlas of the American West* (Chicago: Rand McNally and Co., 1956), p. 6.

37. Matthews-Northrup Co., *Developed Art of Map Making and Made World Market for His Wares* (Buffalo: Walden, Sons, and Mott, 1925).

38. Barbara Bartz Petchenik, "The Twentieth-Century Mapping of the American Revolutionary War," in *Mapping the American Revolutionary War* (Chicago: University of Chicago Press, forthcoming).

39. Quoted in ibid.

40. Richard Edes Harrison to David Woodward, 17 April 1969.

41. Ferdinand von Schwedler to David Woodward, 14 June 1968.

42. M. J. Stanton, "Brief of Rand McNally and Company," in *Congressional Hearings: Tariff Readjustment* (Washington: Government Printing Office, 1929), 7114.

43. Wallace B. Mitchell, Transcript of interview held at Cambridge, Mass., 27 June 1968.

44. John Livesy Ridgway, *The Preparation of Illustrations for Reports of the United States Geological Survey, with Brief Descriptions of Processes of Reproduction* (Washington, D.C.: Government Printing Office, 1920), pp. 54–55. It should be mentioned that, at this time, this method was used only for headings, titles, etc., and not for the smaller names on the main part of the map.

45. Elizabeth M. Harris, "Experimental Graphic Processes in England, 1800–1859," *Journal of the Printing Historical Society* 5 (1969): 41–80.

46. Alfred Dawson and Henry Thomas Dawson, "Improvements in Typographic Etching and Engraving," *Patent Specification no. 1626*, London, British Patent Office, 29 May 1872.

47. Bernard H. Newdigate, "Contemporary Printers. II. Emery Walker," *Fleuron* 4 (1927): 68.

48. *Inland Printer* 55 (1915): 321.

49. John W. Hiltman, "The Question of Mapmaking," *Publishers Weekly* 116 (1929): 2305.

50. Quoted in John W. Hiltman, "As to the Imported Maps," *Publishers Weekly* 116 (1929): 1941.

51. *Geographical Journal* 69 (1927): 592.

52. *Geographical Review* 40 (1950): 678.

53. C. Koeman, "The Application of Photography to Map Printing and the Transition to Offset Lithography," in David Woodward, ed., *Five Centuries of Map Printing* (Chicago: University of Chicago Press, 1975).

54. Stanton, "Brief of Rand McNally and Company."

55. Erwin Raisz, *General Cartography*, 2d ed. (New York: McGraw-Hill, 1948), pp. 50–51.

56. Norman J. W. Thrower, "The County Atlas in the United States," *Surveying and Mapping* 21 (1961): 365–72.

57. Walter W. Ristow, "United States Fire Insurance and Under-

writers' Maps," *Library of Congress Quarterly Journal* 25 (1968): 194–218.

58. Walter W. Ristow, "American Road Maps and Guides," *Scientific Monthly* 62 (1946): 397–406.

CHAPTER THREE

1. H. B. Hatch and A. A. Stewart, *Electrotyping and Stereotyping: A Primer of Information* . . . (Chicago: Committee on Education, United Typothetae of America, 1918), p. 38.

2. Edward Palmer, "Improvements in Producing Printing Surfaces and Printing China, Pottery Ware, Music, Maps, and Portraits," *Patent Specification no. 8987*, London, British Patent Office, 11 December 1841. *Albata* is a synonym for *German silver*, a white alloy of nickel, zinc, and copper.

3. Charles E. Dawson, "Reminiscences of an Old Process Engraver: Wax Engraving," *Inland Printer* 40 (1908): 697–99; *American Dictionary of Printing and Bookmaking* (New York: Howard Lockwood and Co., 1894), p. 366. The use of copper is advised in the following published sources: Benedict (1912), Bormay (1900), Hackleman (1921), Meldau (1947), National Map Company (n.d.), Palmer (1841, 1842, 1843), Reinhold (1886), Schwartz (1930), Servoss (1906), Sherman (1915), and others. The full descriptions of these works are found in the Bibliography.

4. Stanley Larson, Transcript of interview held at Cambridge, Mass., 27 June 1968.

5. Herman Reinhold, "Improved Wax Process," *Inland Printer* 3 (1886), 389–90; Rand McNally and Co. [Russell Voisin], "Wax Engraving as a Medium for Producing Maps," unpublished paper, Rand McNally and Co., Chicago, 1965; Benedict, "Map Engraving"; Ernest Meldau, "Wax Engravings," *Graphic Arts Production Yearbook* 7 (1947) 106; Sherman, "Modern Mapmaking."

6. Mitchell, Interview. Reinhold, "Improved Wax Process," p. 390.

7. For various staining solutions, see Dawson and Dawson, *Patent no. 1626*; Reinhold, "Improved Wax Process"; Sherman, "Modern Mapmaking," p. 322; Robert D. Servoss, "The Wax Process," in Frederick H. Hitchcock, *The Building of a Book* (New York: Grafton Press, 1906); Benedict, "Map Engraving," p. 206. The formula given in "Preparation of Wax for Wax Engraving," *Inland Printer* 16 (1895): 305, is 100 grains of iodide of potassium to 1 oz. water, to be left standing 1 1/2 hours. See also Rand McNally and Co., "Wax Engraving," p. 2; Edward Palmer, "Improvements in Producing Printing and Embossing Surfaces," *Patent Specification no. 9227*, London, British Patent Office, 15 July 1842.

8. Benedict, "Map Engraving," p. 206; Mitchell, Interview; Palmer, *Patent no. 9227*, p. 3; Benjamin Schwartz, "How Wax Engravings Are

Produced, and the Needs They Serve in Modern Printing," *Inland Printer* 85 (1930): 67.

9. Schwartz, "Wax Engravings," p. 67. A. W. Thompson, "Improvement in Electrographic Printing," *Patent Specification no. 4012*, Washington, D.C., U.S. Patent Office, 26 April 1845. National Map Company, "Present Day Methods," paper printed for the internal use of the National Map Company (n.d., ca. 1920), p. 1; R. Landes, "Wax Engraving," unpublished note (date and place of writing unknown, but post-1947); Charles W. Hackleman, *Commercial Engraving and Printing* (Indianapolis: Commercial Engraving Publishing Company, 1921), p. 377. Lewis H. Macomber, Kalamazoo Paraffine Company, to David Woodward, 30 August 1968; Rand McNally and Co., "Wax Engraving," p. 2.

10. Mitchell, Interview.

11. Rand McNally and Co., "Wax Engraving," p. 2.

12. Dawson and Dawson, *Patent no. 1626*; Benedict, "Map Engraving," p. 206; *American Dictionary of Printing and Bookmaking*, p. 366.

13. Mitchell, Interview, p. 2.

14. Meldau, "Wax Engravings," p. 105.

15. Thompson, *Patent no. 4012.*

16. Wallace B. Mitchell to David Woodward, 29 November 1965.

17. Alfred Sidney Johnson, "How Maps and Atlases Are Made," *Publishers Weekly* 101 (1922): 1226–27.

18. Benedict, "Map Engraving," p. 207. Mitchell, Interview, p. 2.

19. Mitchell to Woodward, 29 November 1965. Mitchell, Interview, p. 2: "We were like crows after dentists. We'd go and find any old tools that they had, and if they had a stainless steel one, that was a find."

20. Rand McNally and Co., "Wax Engraving," p. 3.

21. Dawson and Dawson, *Patent no. 1626*, pp. 5–6. Benedict, "Map Engraving," p. 207. Dawson, "Reminiscences." Rand McNally and Co., "Wax Engraving," p. 3. Sherman, "Modern Mapmaking," p. 323.

22. James Shirley Hodson, *An Historical and Practical Guide to Art Illustration, in Connection with Books, Periodicals and General Decoration* (London: Sampson Low, Marston, Searle and Rivington, 1884), p. 150.

23. Servoss, "Wax Process," p. 178.

24. Rand McNally and Co., "Wax Engraving," p. 3.

25. Sherman, "Modern Mapmaking," p. 323.

26. J. Stanley Haas, Transcript of interview held at Buffalo, New York, 18 December 1968; Landes, "Wax Engraving," p. 2.

27. Meldau, "Wax Engravings," p. 105.

28. Mitchell to Woodward, 29 November 1965.

29. Crawford C. Anderson to David Woodward, 28 October 1968.

30. Rand McNally and Co., "Wax Engraving," p. 9.

31. National Map Company, "Present Day Methods"; Rand McNally and Co., "Wax Engraving," p. 9; Charles H. Waite, "Letter-Press Map-Printing," *Patent Specification no. 158,611*, Washington, D.C., U.S. Patent Office, 5 June 1874.

32. Rand McNally and Co., "Wax Engraving," p. 9.

33. Ibid., p. 4.

34. Haas, Interview, p. 4; Landes, "Wax Engraving," p. 2.

35. Mitchell, Interview, p. 3.

36. Reinhold, "Improved Wax Process," p. 389; Haas, Interview, p. 2; L. L. Poates Engraving Co., *Poates Wax Engraving Superiority* (New York: L. L. Poates Engraving Co., 1913), p. 3. Mitchell to Woodward, 29 November 1965, p. 1. J. W. Clement Co., "Notes on Cartography," *Clement Comments* 18 (1935): 6. Ferdinand von Schwedler, Transcript of interview held at Denville, New Jersey, 23 June 1968.

37. Sherman, "Modern Mapmaking," p. 323. A *galley* is a tray (usually about 24 x 7 inches) in which composed type is placed for storage and proofing.

38. Anderson to Woodward, 28 October 1968.

39. Benedict, "Map Engraving," p. 207; Rand McNally and Co., "Wax Engravings," p. 67; Sherman, "Modern Mapmaking," p. 323; Meldau, "Wax Engravings," p. 105; Mitchell to Woodward, 29 November 1965, p. 1. Rand McNally and Co., "Wax Engraving," p. 5. Sherman, "Modern Mapmaking," p. 323. Schwartz, "How Wax Engravings Are Produced," p. 67. Schwartz recommends 85° F.; Dawson and Dawson, *Patent no. 1626*, p. 6, recommend 100°–110° F. Schwartz, "How Wax Engravings Are Produced," p. 67. Rand McNally and Co., "Wax Engraving," p. 5; Mitchell, Interview, pp. 4, 2: "Many stampers would have a callous an eighth of an inch thick on the index finger from fingering-in type."

40. Mitchell, Interview, p. 4. Rand McNally and Co., "Wax Engraving," p. 5.

41. Rand McNally and Co., "Wax Engraving," p. 4.

42. Dawson, "Reminiscences"; Benedict, "Map Engraving," p. 207; Meldau, "Wax Engravings," p. 105; Hackleman, *Commercial Engraving*, p. 379. Dawson and Dawson, *Patent no. 1626*, p. 6.

43. Sherman, "Modern Mapmaking," p. 324. Rand McNally and Co., "Wax Engraving," p. 4.

44. Rand McNally and Co., "Wax Engraving," p. 2.

45. Mitchell, Interview, p. 2.

46. Hackleman, *Commercial Engraving*, p. 107; Landes, "Wax Engraving," p. 3; Mitchell, Interview, p. 3; Palmer, *Patent no. 9227*, p. 3.

47. Dawson, "Reminiscences."

48. Dawson and Dawson, *Patent no. 1626*, p. 8.

49. Dawson, "Reminiscences," p. 698.

50. Mitchell, Interview, p. 2.

51. Thompson, *Patent no. 4012.* Meldau, "Wax Engravings," p. 106.

52. Rand McNally and Co., "Wax Engraving," p. 7. F. J. Henry, "The Preparation of Wax Molds for the Battery," *Inland Printer* 17 (1896): 279.

53. Hatch and Stewart, *Electrotyping.* E. Stanley Haas, Interview.

54. Henry, "Wax Molds," pp. 280, 279. Landes, "Wax Engraving," p. 4.

55. Hatch and Stewart, *Electrotyping,* p. 4.

56. Landes, "Wax Engraving," p. 4; Meldau, "Wax Engravings," p. 106.

57. Mitchell, Interview, p. 3.

58. Rand McNally and Co., "Wax Engraving," p. 7; National Map Company, "Present Day Methods," p. 2; Meldau, "Wax Engravings," p. 107; von Schwedler, Interview, p. 2.

59. Von Schwedler, Interview, p. 2: "The finisher at Hammond's was 76 years old, and as soon as he died, we couldn't use any more of our letterpress plates."

60. Rand McNally and Co., "Wax Engraving," p. 10.

61. Mitchell, Interview, p. 2.

62. Rand McNally and Co., "Wax Engraving," p. 11. Von Schwedler, Interview, p. 1. Rand McNally and Co., "Wax Engraving," p. 11.

63. For an example of a map made with three halftone color plates and one wax-engraved black plate see *Graphic Arts and Crafts Year Book* 5 (1911–12): 256.

CHAPTER FOUR

1. R. B. Fishenden, "The Artist and the Printer," in *Printing in the Twentieth Century* (London: The Times Publishing Company, 1930), pp. 98–99.

2. For a further elaboration of this theme, see Arthur H. Robinson, "Mapmaking and Map Printing: The Evolution of a Working Relationship," in David Woodward, ed., *Five Centuries of Map Printing* (Chicago: University of Chicago Press, 1975).

3. Mitchell, Interview.

4. Stanton, "Brief of Rand McNally and Company."

5. Rand McNally and Co., "Wax Engraving."

6. Von Schwedler, Interview.

7. Larson, Interview.

8. Mitchell, Interview.

9. Von Schwedler, Interview.

10. Ibid.

CHAPTER FIVE

1. Von Schwedler, Interview.

2. Although the French painters Daubigny and Corot experimented with scribing in 1854 (J. J. Klawe, "Strip Mask Techniques," in *Proceed-*

ings of the Cartographic Symposium [Edinburgh, 1962]), the earliest technical description yet found is in John Jacob Holtzapffel's *Turning and Mechanical Manipulation* (London: Holtzapffel and Co., 1884). See David Woodward, "A Note on the History of Scribing," *Cartographic Journal* 3 (1966): 58.

3. Rand McNally and Co., "Wax Engraving," p. 2.

4. Ibid., p. 9.

5. Ibid., p. 4. See above, p. 62, for description of method.

6. Beatrice Warde, "Printing Should Be Invisible," in Paul Bennett, ed., *Books and Printing* (Cleveland: World Publishing Company, 1963), pp. 109–14.

7. Charles Eckstein, "New Method for Reproducing Maps and Drawings," paper printed for private circulation only at the International Exhibition at Philadelphia, 1876.

8. Barbara Bartz [Petchenik], "Type Variation and the Problem of Cartographic Type Legibility," *Journal of Typographic Research* 3 (1969): 127–44.

9. Eduard Imhof, "Die Anordnung der Namen in der Karte," *Internationales Jahrbuch für Kartographie* 2 (1962): 93–129.

10. Rand McNally and Co., "Wax Engraving," p. 5.

11. Three-point type, known as Brilliant, is dismissed as a "curiosity" in W. W. Pasko, *American Dictionary of Printing and Bookmaking* (New York: Howard Lockwood and Co., 1894), p. 70.

12. For example, the half nonpareil (3-point) type cast by Henri Didot in 1825. It was reported that no type of this size had been cast by American type founders by 1894 (*American Dictionary of Printing and Bookmaking*, pp. 377–78).

13. Hodson, *An Historical and Practical Guide*, p. 152.

14. Rand McNally and Co., "Wax Engraving," p. 4.

15. Matthews-Northrup Co., *Developed Art of Map Making*, p. 5.

16. It should be pointed out that when wax-engraved plates were transferred to lithographic plates, as was frequently the case, such impression characteristics as indentation, sharpness, and density did not apply.

17. See A. G. Fegert, "All Processes Required," *Inland Printer* 96 (1935): 21–27.

18. Ibid., p. 24.

19. Larson, pp. 1, 2.

20. This term is borrowed from Raymond Lister, *How to Identify Old Maps and Globes* (London: G. Bell and Sons, 1965), p. 52.

21. Sidney Edwards Morse, "Cerography," *New York Observer* 17 (20 July 1839): 114, col. 6. The Napier press is believed to be the earliest power cylinder press imported to the United States. It was imitated by early American printing-machine manufacturers such as Hoe, Taylor, and so forth. See W. Turner Berry and H. Edmund Poole, *Annals of Printing* (London: Blandford Press, 1966), p. 220.

22. Sidney Edwards Morse, Preface, in *Cerographic Atlas of the United States* (New York: S. E. Morse and Co., 1842).

23. George Ehrlich, "Technology and the Artist: A Study of the Interaction of Technological Growth and Nineteenth Century American Pictorial Art" (Ph.D. diss., University of Illinois, Urbana, 1960), p. 18.

24. Berry and Poole, *Annals of Printing*, p. 196.

25. Ibid., p. 230.

26. Waite, *Patent no. 158,611.*

27. J. Stanley Haas to David Woodward, 24 June 1969.

28. Rand McNally and Co., "Wax Engraving," p. 11.

29. Morse, "Cerography."

30. Rand McNally and Co., "Wax Engraving," p. 4.

31. *Graphic Arts Production Yearbook* 6 (1940): 85.

32. Morse, "Cerography."

33. Duncan M. Fitchet, "100 Years and Rand McNally," *Surveying and Mapping* 16 (1956): 126–32.

34. Rand McNally and Co., "Wax Engraving," p. 10.

35. Mitchell, Interview.

CHAPTER SIX

1. Stanley Larson, Interview.

2. Erwin Raisz, *General Cartography*, 2d ed. (New York: McGraw-Hill, 1948), p. 50.

Bibliography

Adams, Cyrus. *Bulletin of the American Geographical Society* 38 (1906): 123–25.

———. "Maps and Map Making." *Bulletin of the American Geographical Society* 44 (1912): 194–201.

American Dictionary of Printing and Bookmaking. New York: Howard Lockwood and Co., 1894.

Anderson, Crawford C. "How an Atlas Is Made." *Clement Comments* 20 (1937): 10–12.

Antonelli, Michael. "The Role of Maps in Early American Geographies, 1784–1890." M.A. thesis, Syracuse University, 1968.

———. "The Role of Maps in Early American Geographies." *Proceedings of the Ninth Annual Meeting of the New York–New Jersey Division, Association of American Geographers* 2 (1969): 63–78.

Arenson, Saul B. "Relief Printing." *Journal of Chemical Education* 7 (1930): 1632–41.

"As to Wax or Cereographic Engraving." *Inland Printer* 39 (1907): 713.

Atkins, William. *The Art and Practice of Printing.* Vol. 4, *Photo-Engraving, Electrotyping and Stereotyping.* London: Isaac Pitman, 1934.

Bacon, George Washington. *Patent Specification no. 57056,* London, British Patent Office, 7 August 1866.

Bagrow, Leo. *History of Cartography,* revised by R. A. Skelton. London: C. A. Watts, 1964.

Bartz, Barbara S. "Type Variation and the Problem of Cartographic Type Legibility." *Journal of Typographic Research* 3 (1969): 127–44; 3 (1969): 387–98; 4 (1970): 147–67. *See also* Petchenik, Barbara Bartz.

Bay, Helmuth. "The Beginning of Modern Road Maps in the United States." *Surveying and Mapping* 12 (1952): 413–16.

Bibliography

————. *The History and Technique of Map Making.* New York: New York Public Library, 1943.

Benedict, George H. "Map Engraving." *Printing Art* 19 (1912): 205–9.

Berry, W. Turner, and Poole, H. Edmund. *Annals of Printing.* London: Blandford Press, 1966.

Bigmore, E. C., and Wyman, C. W. H. *A Bibliography of Printing with Notes and Illustrations.* London: Bernard Quaritch, 1880.

Blaikie, W. B. "How Maps Are Made." *Scottish Geographical Magazine* 7 (1891): 419–34.

Blum, William, and Hogaboom, George B. *Principles of Electroplating and Electroforming.* New York: McGraw-Hill, 1924.

Bormay, W. J. "Wax Engraving." *Inland Printer* 26 (October 1900): 85–86.

Bosse, Heinz. "Kartentechnik,—II Vervielfältigungsverfahren." In *Ergänzungsheft Nr. 245 zu Petermanns Geographische Mitteilungen.* Gotha: Justus Perthes, 1951.

Breitkopf, Johann Gottlieb Immanuel. *Über den Druck der Geographischen Charten, Nebst beygefügter Probe einer durch die Buchdruckerkunst gesetzen und gedruckten Landcharte.* Leipzig: published by the author, 1777.

Brigham, Albert Perry, and Dodge, Richard E. "Nineteenth Century Textbooks of Geography." In *The Teaching of Geography*, National Society for the Study of Education, 32d Yearbook. Bloomington, Ill.: Public School Publishing Co., 1933.

Brown, John Howard, ed. *Lamb's Biographical Dictionary of the United States.* Boston: Federal Book Company of Boston, 1903.

Brown, Lloyd A. *The Story of Maps.* Boston: Little, Brown and Co., 1949.

Brown, Ralph H. "The American Geographies of Jedidiah Morse." Association of American Geographers, *Annals* 31 (1941): 145–217.

Burkhart, C. A. "Map Making: A Sketch." *Education* 58 (1938): 271–78.

Byrn, Edward W. *The Progress of Invention in the Nineteenth Century.* New York: Munn and Co., 1900.

Carpenter, Charles. *History of American Schoolbooks.* Philadelphia: University of Pennsylvania Press, 1963.

Carpenter, J. "Concerning the Graphotype." *Once a Week* 3 (1867): 181–84.

Carr, J. Comyns. "Book Illustration, Old and New." *Journal of the Society of Arts* 30 (1882): 1035–43, 1045–53, 1055–62.

Cave, Roderick, and Wakeman, Geoffrey. *Typographia Naturalis.* Wymondham, England: Brewhouse Press, 1967.

Bibliography

"Cereographic or Wax Plate Engraving." *Inland Printer* 38 (1906): 404.

"Cereography or Wax Engraving." *Inland Printer* 29 (1902): 422.

"Cerographic Process versus Lithography." *Inland Printer* 16 (January 1896): 425.

"Cerographic, or Wax Engraving Process." *Inland Printer* 52 (1913): 390.

"Cerographs or Engraving in Wax." *Inland Printer* 25 (1900): 224.

Chatto, William Andrew. *A Treatise on Wood Engraving, Historical and Practical.* London: Charles Knight and Co., 1861.

Clement Co., J. W. *The Making of Fine Maps.* Buffalo, N.Y.: J. W. Clement Co., 1927.

———. "Notes on Cartography." *Clement Comments* 18 (1935): 5–6.

Clow, H. B. "Why a Tariff on Maps?" *Publishers Weekly* 116 (1929): 2161–64.

Cowan, T. W. *Wax Craft: All about Beeswax.* London: Sampson Low, 1908.

Cram, George F., Co. Typewritten notes concerning the history of the company (no date).

Crerar Library, John. *A List of Books on Industrial Arts.* Chicago: Trustees of John Crerar Library, 1904.

Crone, Gerald Roe. "The Evolution of Map Printing." In *Printing: A Supplement Published by "The Times" on the Occasion of the 10th International Printing Machinery and Allied Trades Exhibition.* 1955.

———. *Maps and Their Makers.* London: Hutchinson University Library, 1964.

Crosthwaite, W. H. "A Visit to Some Map-Printing Works in England." *Cairo Scientific Journal* 5 (1911): 290–94.

Culler, Ned. "The Development of American Geography Textbooks from 1840 to 1890." Ed.D. diss., University of Pittsburgh, 1945.

Dahl and Sinnott (Wood Engravers). "Wood Engraving." *Penrose's Pictorial Annual: The Process Year Book* 19 (1913–14): 149–54.

Dawson, Alfred, and Dawson, Henry Thomas. "Improvements in Typographic Etching and Engraving." *Patent Specification no. 1626,* London, British Patent Office, 29 May 1872.

Dawson, Charles E. "Reminiscences of an Old Process Engraver: Wax Engraving." *Inland Printer* 40 (1908): 697–99.

Dembour, A. *Description d'un nouveau procédé de gravure en relief sur cuivre, dite ectypographie, inventé par A. Dembour.* Metz: S. Lamort, 1835.

Dowgray, John Gray Laird. "A History of Harper's Literary Magazines, 1850–1900." Ph.D. diss., University of Wisconsin, 1955.

Dryer, Charles Redway. "A Century of Geographic Education in the

Bibliography

United States." Association of American Geographers, *Annals* 14 (1924): 117–49.

Dubois, Arthur L. "Map Drafting and Reproduction." *Graphic Science* 3 (1961): 18–21.

"Duty on Maps." *Publishers Weekly* 116 (1929): 1338–39.

Eckert, Max. *Die Kartenwissenschaft.* Berlin and Leipzig: Vereinigung Wissenschaftlicher Verleger, Walter de Gruyter and Co., 1921.

Eckstein, Charles. "New Method for Reproducing Maps and Drawings." Paper printed for private circulation only, International Exhibition at Philadelphia, 1876.

Ehrlich, George. "Technology and the Artist: A Study of the Interaction of Technological Growth and Nineteenth Century American Pictorial Art." Ph.D. diss., University of Illinois, Urbana, 1960.

Eldridge, Albert G.; Abrams, Alfred W.; Jansen, William; and Shryock, Clara M. "Maps and Map Standards." In *The Teaching of Geography.* Bloomington, Ill.: Public School Publishing Co., 1933.

"Etching in Wax Engraving." *Inland Printer* 26 (1900): 278.

Exman, Eugene. *The Brothers Harper.* New York: Harper and Row, 1965.

Fegert, A. G. "All Processes Required." *Inland Printer* 96 (1935): 21–27; 97 (1935): 68.

Fielding, Theodore Henry Adolphus. *The Art of Engraving, with the Various Modes of Operation.* London: Ackerman and Co., 1841.

"First Halftone Printed in N.Y." *Editor and Publisher* 73 (1940): 24.

Fischer, Ernst. "Zweihundert Jahre Naturselbstdruck." In *Gutenberg Jahrbuch.* Mainz: Gutenberg-gesellschaft, 1933.

Fishenden, R. B. "The Artist and the Printer." In *Printing in the Twentieth Century.* London: The Times Publishing Co., 1930.

Fitchet, Duncan M. "100 Years and Rand McNally." *Surveying and Mapping* 16 (1956): 126–32.

"Formula for Wax Ground for Wax Engraving." *Inland Printer* 16 (October 1895): 73.

Gates, Paul W. *The Illinois Central Rail-Road and Its Colonization Work.* Cambridge, Mass.: Harvard University Press, 1934.

Goode, J. Paul. "The Map as a Record of Progress in Geography." Association of American Geographers, *Annals* 17 (1927): 1–14.

Gore, George. *Theory and Practice of Electro-Deposition.* London: Houlston and Stoneman, 1856.

Graphic Arts and Crafts Year Book 5 (1911–12).

Graphic Arts Production Yearbook 3 (1936): 120; 6 (1940): 85, 204.

"Graphotype." *Nature and Art* 1 (1866): 219–22.

Bibliography

Gress, Edmund G. *Fashions in American Typography, 1780-1930.* New York: Harper and Brothers, 1931.

Grolier Club. *Type Specimen Books and Broadsides Printed before 1900 Exhibited at the Grolier Club on November the Eighteenth 1926.* New York: Grolier Club, 1926.

Haas, J. Stanley. Transcript of interview held at Buffalo, N.Y., 18 December 1968.

Haas, William. "The Wax Process." Typewritten manuscript of an address given to fellow printers (date unknown, ca. 1920).

Hackleman, Charles W. *Commercial Engraving and Printing.* Indianapolis, Ind.: Commercial Engraving Publishing Company, 1921.

Hamer, Philip M., ed. *A Guide to Archives and Manuscripts in the United States.* New Haven, Conn.: Yale University Press, 1961.

Harper, J. Henry. *The House of Harper: A Century of Publishing in Franklin Square.* New York: Harper and Brothers, 1912.

Harper and Brothers. Manuscripts relating to contracts between Sidney Edwards Morse, Samuel A. Breese, Henry A. Munson and the firm of Harper & Brothers, 1844-46. Archives, Harper and Row.

Harris, Elizabeth M. "Experimental Graphic Processes in England, 1800-1859." *Journal of the Printing Historical Society* 4 (1968): 33-86; 5 (1969): 41-80; 6 (1970): 53-89.

Harrison, James. *Printing Patents: Abridgements of Patent Specifications Relating to Printing, 1617-1857.* London: Printing Historical Society, 1969.

Harvard University Library. *Bibliography and Bibliography Periodicals.* Cambridge, Mass.: Harvard University Press, 1966.

Hatch, Harris B., and Stewart, A. A. *Electrotyping and Stereotyping: A Primer of Information* . . . Chicago: Committee on Education, United Typothetae of America, 1918.

Henry, F. J. "The Preparation of Wax Molds for the Battery." *Inland Printer* 17 (1896): 279-80.

Hentschel, Carl. *How to Print Carl Hentschel Colortype Blocks.* London: privately printed, 1900.

Hiltman, John W. "As to the Imported Maps." *Publishers Weekly* 116 (1929): 1940-41.

———. "The Question of Map Making." *Publishers Weekly* 116 (1929): 2305.

Hind, Arthur Mayger. *A History of Engraving and Etching from the 15th Century to the year 1914.* New York: Houghton Mifflin Co., 1923.

Hinks, Arthur R. "The Science and Art of Map-making." *Scottish*

Bibliography

Geographical Magazine 4 (1925): 321–36.

Hitchcock, Frederick H. *The Building of a Book.* New York: Grafton Press, 1906.

Hodson, James Shirley. "Automatic Engraving: Technical and Historical." *Inland Printer* 5 (1888): 473–76; 6 (1889): 285–88.

——. *An Historical and Practical Guide to Art Illustration, in Connection with Books, Periodicals, and General Decoration.* London: Sampson Low, Marston, Searle and Rivington, 1884.

"How Wax Engravings Are Produced." *Inland Printer* 85 (1930): 67–68.

"Illustrated Newspapers." *Inland Printer* 1 (1884): 16.

Imhof, Eduard. "Die Anordnung der Namen in der Karte." *Internationales Jahrbuch für Kartographie* 2 (1962): 93–129.

"The Improved Hammond Globes." *Publishers Weekly* 117 (1930): 1135.

Industrial Arts Index. New York: H. W. Wilson Co., annually.

International Association of Printing House Craftsmen. *Index to Graphic Arts Periodical Literature.* (See Karch, R. Randolph.)

International Printing Machinery and Allied Trades Exhibition (Eleventh). *Printing and the Mind of Man: Catalogue of the Exhibition at the British Museum and at Earls Court, London, 16–27 July 1963.* London: F. W. Bridges and Sons and the Association of British Manufacturers of Printers' Machinery (Proprietary) Ltd., 1963.

Ireland, Oscar B. *Bulletin of the American Geographical Society* 32 (1900): 168–69.

Ivins, William Mills, Jr. *Prints and Visual Communication.* Cambridge, Mass.: Harvard University Press, 1953.

Jewkes, John; Sawers, David; and Stillerman, Richard. *The Sources of Invention.* London: Macmillan and Co., 1958.

Joerg, W. L. G. "Post-War Atlases: A Review." *Geographical Review* 13 (1923): 598.

Johnson, Alfred Sidney. "How Maps and Atlases Are Made." *Publishers Weekly* 101 (1922): 1033–34, 1102–4, 1166–69, 1226–28.

Karch, R. Randolph. *How to Recognize Type Faces.* Bloomington, Ill.: McKnight and McKnight, 1952.

——. *Index to Graphic Arts Periodical Literature.* Evanston, Ill.: Education Commission, International Association of Printing House Craftsmen, 1942.

——. "Stereotyping and Wax Engraving." *Industrial Arts and Vocational Education* 23 (1934): 12.

Kaser, David. *The Cost Book of Carey & Lea, 1825–1838.* Philadelphia: University of Pennsylvania Press, 1963.

154

Bibliography

Keates, John S. "The History of Cartography." *Cartographic Journal* 1 (1964): 51.

Kinniburgh, Ian A. G. "Cartographic Essays Reprinted." *Cartographic Journal* 4 (1967): 140.

Kirk, Albert. *A Selected List of Graphic Arts Literature, Books and Periodicals.* London: British Federation of Master Printers, 1948.

Knight, Charles. *The British Mechanic's and Labourer's Hand Book, and True Guide to the United States, with Ample Notices Respecting Various Trades and Professions.* London: Charles Knight and Co., 1840.

Koehler, S. "Typographic Etchings." *American Art Review* 1 (1880): 222–23.

Landes, R. "Wax Engraving." Unpublished note, date and place of writing unknown, but post-1947; follows Haas (ca. 1920) in parts.

Larson, Stanley. Transcript of interview held at Cambridge, Mass., 27 June 1968.

LeGear, Clara E. *A List of Geographical Atlases in the Library of Congress.* Vols. 5 and 6. Washington, D.C.: Government Printing Office, 1958, 1963. For vols. 1–4 see Phillips, Philip Lee.

Legros, Lucien Alphonse, and Grant, John Cameron. *Typographical Printing-Surfaces: The Technology and Mechanism of Their Production.* London: Longmans, Green and Co., 1916.

Lister, Raymond. *How to Identify Old Maps and Globes.* London: G. Bell and Sons, 1965.

Lord, H. D. *Memorial of the Family of Morse, Compiled from the Original Records for the Hon. Asa Porter Morse.* Cambridgeport, Mass., 1896.

"A Loss to Publishing." *Publishers Weekly* 116 (1929): 2926.

Mabee, Carleton. *The American Leonardo: The Life of Samuel F. B. Morse.* New York: Alfred A. Knopf, 1943.

Mabley, William Tudor. "Certain Improvements to Be Used in Printing, Embossing, or Impressing." *Patent Specification no. 8743*, London, British Patent Office, 17 December 1840.

McElheran, John. "Feed-Motion for Cerotypography." *Patent Specification no. 21208*, Washington, D.C., U.S. Patent Office, 17 August 1858.

———. "Improvement in Graphotype." *Patent Specification no. 19707*, Washington, D.C., U.S. Patent Office, 23 March 1858.

McNally, Andrew, III. *The World of Rand McNally.* New York: Newcomen Society of America, 1956.

Mathiot, George. "On the Electrotyping Operations of the U.S. Coast

Survey." *American Journal of Science and Arts* 15 (1853): 305–9.

Matthews-Northrup Co. *Developed Art of Map Making and Made World Market for His Wares.* Buffalo, N.Y.: Walden, Sons, and Mott, 1925.

Mayer, Max. "Maps and Their Making." *Publishers Weekly* 117 (1930): 2841–47; 118 (1930): 70–78, 973–76, 1663–66.

Meldau, Ernest. "Wax Engravings." *Graphic Arts Production Year Book* 7 (1947): 105–7.

Mertle, J. S. "Wax Engraving." *Graphic Arts Monthly* 11 (1939): 41–44.

Mills, George J. *Sources of Information in the American Graphic Arts.* Pittsburgh: Carnegie Institute of Technology, 1951.

Mitchell, Wallace B. Transcript of interview held at Cambridge, Mass., 26 June 1968.

Mol, A. "Modern Mapmaking." *Inland Printer* 55 (1915): 486.

Morse, Edward Lind. *Samuel F. B. Morse.* Boston, 1914.

Morse, John Howard. *Morse Genealogy. . . .* New York: Springfield Printing and Binding Company, 1903–5.

Morse, Richard Cary. Correspondence relative to cerography. Morse Family Papers, Yale University Library, New Haven, Conn. To Sidney Edwards Morse, New York, 18 September 1845; New York, 22 October 1845. To Samuel Finley Breese Morse, New York, 1 August 1848.

Morse, Sidney Edwards. "Cerography." *New York Observer* 17 (13 July 1839): 110, col. 4.

―――. "Cerography." *New York Observer* 17 (20 July 1839): 114.

―――"Cerography." *New York Observer* 17 (19 October 1839); p. 166, col. 2.

―――. Correspondence relative to cerography. Morse Family Papers, Yale University Library, New Haven, Conn. To Richard Cary Morse, New York, 29 January 1841; Hook of Holland (7 August 1845); London (10 October 1845); Glasgow (17 July 1846); Frankfurt (28 September 1846) to *The Times:* London (18 May, 1847). For notes about these and other of Morse's letters, see *Yale University Library Gazette* 23 (1948): 151–54.

―――. Editorial. *New York Observer* 18 (4 January 1840).

―――. Editorial. *New York Observer* 19 (2 January 1841).

―――. "Manufacture of Plates for Printing or Embossing." *Patent Specification no. 12022*, London, British Patent Office, 13 January 1848.

―――. "Maps: New Mode of Engraving." *New York Observer* 17

Bibliography

(29 June 1839): 105.

―――. *Memorabilia in the Life of Jedidiah Morse, D. D., Former Pastor of the First Church in Charlestown, Mass.* Boston, 1867.

―――. Preface. In *The Cerographic Atlas of the United States.* New York: S. E. Morse and Co., 1841. (Also in *New York Observer* 20 [12 February 1842]: 26.)

Mott, Frank Luther. *American Magazines, 1865-1880.* Iowa City: The Midland Press, 1928.

―――. *History of American Magazines.* Cambridge, Mass.: Harvard University Press, 1938-57.

Motte, Philip H. de la. *On the Various Applications of Anastatic Printing and Papyrography with Illustrative Examples.* London: David Bogue, 1849.

Mulliner, Beulah A. "The Development of Physiography in American Textbooks." *Journal of Geography* 11 (1912): 319-24.

National Map Company. "Present Day Methods." Paper printed for the internal use of the National Map Company (no date, ca. 1920).

Nebenzahl, Kenneth. "A Stone Thrown at the Map Maker." *Papers of the Bibliographical Society of America* 55 (1961): 283-88.

"New Atlases in Demand." *Publishers Weekly* 118 (1930): 2198.

Newdigate, Bernard H. "Contemporary Printers. II. Emery Walker." *Fleuron* 4 (1926): 63-69.

Nietz, John Alfred. *Old Textbooks.* Pittsburgh: University of Pittsburgh Press, 1961.

Nordenskiold, Eric. *History of Biology.* New York: Tudor, 1935.

Oldfield, Margaret. "Five Centuries of Map Printing." *British Printer* 63 (1955): 69-74.

"100-Year Production Figure: Six Billion Maps and Globes." *Printing Equipment Engineer* 87 (1957): 40-41.

"On the Application of Photography to Printing." *Harper's New Monthly Magazine* 13 (1856): 429-41.

Palmer, Edward. "Glyphography." *Practical Mechanic and Engineer's Magazine* 2 (1843): 468-69.

―――. *Glyphography; or, Engraved Drawing.* . . . London: privately printed, 1843. 3d ed., 1844.

――. "Improvements in Producing Printing and Embossing Surfaces." *Patent Specification no. 9227,* London, British Patent Office, 15 July 1842.

――. "Improvements in Producing Printing Surfaces and Printing China, Pottery Ware, Music, Maps, and Portraits." *Patent Specification no. 8987,* London, British Patent Office, 11 December 1841.

Bibliography

Peddie, Robert Alexander. *Catalogue of Works on Practical Printing, Processes of Illustration and Bookbinding Published since 1900.* London: St. Bride Foundation, 1911.

Petchenik, Barbara Bartz. "The Twentieth-Century Mapping of the American Revolutionary War." In *Mapping the American Revolutionary War.* Chicago: University of Chicago Press, forthcoming.

Pettit, Joseph S. *Modern Reproductive Graphic Processes.* New York: D. van Nostrand, 1884.

Phillips, Philip Lee. *A List of Geographical Atlases in the Library of Congress, with Bibliographical Notes.* Vols. 1–4. Washington, D.C., Government Printing Office, 1909–20. *For vols. 5 and 6 see* LeGear, Clara E.

Poates Wax Engraving Superiority. New York: L. L. Poates Engraving Co., 1913.

"Preparation of Wax for Wax Engraving." *Inland Printer* 16 (December 1895): 305.

"Printing Methods Detector." *Sixth Annual Advertising and Publishing Production Yearbook,* pp. 204–5. New York: Colton Press, 1940.

"The Rage for Illustrations by Cheap, Unsightly, and Unmeaning Cuts...." *Inland Printer* 3 (1885): 99.

Raisz, Erwin. *General Cartography.* 2d. ed. New York: McGraw-Hill, 1948.

———. "Outline of the History of American Cartography." *Isis* 26 (1937): 373–91.

Rand McNally and Co. Early history of the company (manuscript, date unknown).

———. "First Maps" (manuscript), p. 3.

———. *Pioneer Atlas of the American West.* Chicago: Rand McNally and Co., 1956.

———. *Railway Guide,* October 1872, p. xviii; December 1872, p. xxii; April 1873, p. xv.

——— [Russell Voisin]. "Wax Engraving as a Medium for Producing Maps." Unpublished paper. Chicago, 1965.

Rapid Electrotype Co. *From Xylographs to Lead Molds.* Cincinnati: Rapid Electrotype Co., 1921.

Reinhold, Herman. "Improved Wax Process." *Inland Printer* 3 (1886): 389–90.

"A Review of Engraving Methods and Processes." *Graphic Arts Year Book* 3 (1909): 39–48.

Ridgway, John Livesy. *The Preparation of Illustrations for Reports of the United States Geological Survey, with Brief Descriptions of*

Processes of Reproduction. Washington, D.C.: Government Printing Office, 1920.

Ringwalt, J. Luther. "A New System of Engraving Plates for Typographic Presses." *Journal of the Franklin Institute* 97 (1874): 39.

Ristow, Walter W. "American Road Maps and Guides." *Scientific Monthly* 62 (1946): 397–406.

———. "Historical Cartography in the United States, 1959–1963." *Imago Mundi* 17 (1963): 106–114.

———. "United States Fire Insurance and Underwriters' Maps." *Library of Congress Quarterly Journal* 25 (1968): 194–218.

Robinson, Arthur H. "The 1837 Maps of Henry Drury Harness." *Geographical Journal* 121 (1955): 440–50.

———. *Elements of Cartography.* 2d ed. New York: John Wiley, 1960.

———. *The Look of Maps: An Examination of Cartographic Design.* Madison: University of Wisconsin Press, 1952.

Royal Electrotype Company. *Wax Engraving Perfected by Royal.* Philadelphia: Royal Electrotype Co., 1924.

Sahli, John R. "An Analysis of Early American Geography Textbooks from 1784 to 1840." Ph.D. diss., University of Pittsburgh, 1941.

St. Bride Foundation. *Catalogue of the Technical Reference Library.* London: St. Bride Foundation, 1919.

Sampson, Thomas. *Electrotint; or, The Art of Making Paintings in Such a Manner That Copper Plates and "Blocks" Can Be Taken from Them by Means of Voltaic Electricity.* London: Edward Palmer, 1842.

"Scales and Stars." *Inland Printer* 99 (1937): 37–40. A short biography of George H. Benedict.

Schwartz, Benjamin. "How Wax Engravings Are Produced, and the Needs They Serve in Modern Printing." *Inland Printer* 85 (1930): 67–68.

Schwedler, Ferdinand von. Transcript of interview held at Denville, New Jersey, 23 June 1968.

Servoss, Robert D. "The Wax Process." In Frederick H. Hitchcock, *The Building of a Book.* New York: Grafton Press, 1906.

Sherman, George. "Outgrowths of Letterpress, no. 8: Modern Mapmaking." *Inland Printer* 55 (1915): 321–27.

"Sidney Edwards Morse" (obituary). *New York Observer* 49 (28 December 1871): 410.

Silliman, Benjamin Jr. "Electrography or the Electrotype." *American Journal of Science and Arts* 40 (1841): 157–64.

Smith, C. S. "Reflections on Technology and the Decorative Arts in the Nineteenth Century." In Ian M. Quimby and P. A. Earl, eds.,

Bibliography

Technological Innovation and the Decorative Arts. Winterthur Conference Report, 1973. Charlottesville, Va.: University Press of Virginia, 1974.

"Some Railway Map-Making." *Inland Printer* 15 (1895): 500.

Spencer, Thomas. *An Account of Some Experiments Made for the Purpose of Ascertaining How Far Voltaic Electricity May Be Usefully Applied to the Purpose of Working in Metal.* Liverpool: Mitchell, Heaton and Mitchell, 1839.

―――. *Instructions for the Multiplication of Works of Art in Metal, by Voltaic Electricity, with an Introductory Chapter on Electro-Chemical Decompositions by Feeble Currents.* Glasgow: Richard Griffin and Co.; and London: Thomas Tegg, 1840.

Stanton, M. J. "Brief of Rand McNally and Company." In *Congressional Hearings: Tariff Readjustment* (Washington, D.C.: Government Printing Office, 1929), pp. 7112–16.

"The Star Engraving Plates." *Inland Printer* 4 (1886–87): 703.

Stebbins, Lucius. "Improvement in the Mode of Coloring Maps." *Patent Specification no. 1510*, Washington, D.C., U.S. Patent Office, 12 March 1840.

Steeves, H. Alan. *Index to Graphic Arts Printing Processes.* New York: New York Public Library, 1943.

Steinberg, S. H. *Five Hundred Years of Printing.* 2d. ed. Harmondsworth, Middlesex: Penguin Book, 1961.

Taylor, George Rogers. *The Transportation Revolution, 1815–1860.* New York: Harper and Row, 1968.

The Thomas Register. 58th ed. New York: Thomas Publishing Co., 1968.

Thompson, A. W. "Improvement in Electrographic Printing." *Patent Specification no. 4012*, Washington, D.C., U.S. Patent Office, 26 April 1845.

Thompson, Edmund Burke. *Maps of Connecticut for the Years of Industrial Revolution, 1801–1860.* Windham, Conn.: Hawthorn House, 1942.

Thrower, Norman J. W. "The County Atlas in the United States." *Surveying and Mapping* 21 (1961): 365–72.

Timperley, C. H. *A Dictionary of Printers and Printing, with the Progress of Literature, Ancient and Modern, Bibliographical Illustrations, etc., etc.* London: H. Johnson, 1839.

"Tools for Engraving on Wax." *Inland Printer* 15 (June 1895): 312.

Turrell, Edmund. "On a Mode of Preparing Etching-ground for Engravers." *Journal of the Franklin Institute* 2 (1826): 83.

Bibliography

Waite, Charles H. "Letter-Press Map-Printing." *Patent Specification no. 158,611,* Washington, D.C., U.S. Patent Office, 12 September 1875.

Wakeman, Geoffrey. *Victorian Book Illustration, the Technical Revolution.* Detroit: Gale Research Co., 1973.

Walden, Charles C. *Printing Year Book and Almanac.* New York: Walden, Sons, and Mott, 1947.

Walker, Charles V. *Manipulations in the Scientific Arts: Electrotype Manipulation, Part II.* 19th ed. London: George Knight and Sons, 1855.

Warde, Beatrice. "Printing Should Be Invisible." In Paul Bennett, ed., *Books and Printing.* Cleveland: World Publishing Company, 1963.

"Wax Engraving's Beginnings." *Inland Printer* 73 (1924): 593–94.

"Wax Process of Engraving." *Inland Printer* 68 (1922): 650.

"Who Will Sell Maps?" *Publishers Weekly* 119 (1931): 2310.

Wilson, F. J. *Stereotyping and Electrotyping.* 6th ed. London, 1898.

Wood, H. Trueman. *Modern Methods of Illustrating Books.* 3d ed. London: Elliot Stock, 1890.

Woodward, David. "A Note on the History of Scribing." *Cartographic Journal* 3 (1966): 58.

———. *The Use of Type on the Cerographic Maps of Sidney Edwards Morse: A Review of the Evidence.* Madison, University of Wisconsin, 1967.

Woodward, David, ed. *Five Centuries of Map Printing.* Chicago: University of Chicago Press, 1975.

Woodward, David, and Robinson, Arthur H. "Notes on a Genealogical Chart of Some American Commercial Atlas Producers." Special Libraries Association, Geography and Map Division, *Bulletin* 79 (1970): 2–6.

Woodworth, H. H. "Now Is the Time to Sell Atlases." *Publishers Weekly* 119 (1931): 835–38.

Wright, John K. "Highlights in American Cartography, 1939–1949." In *Comte rendu du XVIe congrès international de géographie.* Lisbon, 1959.

Index

Index

Index

Stereotyping, 12
Stylus: for freehand lines, 98, 101; sinuosity characteristics, 100

Tanner, Henry S., 48
Tones: "dotted," 62; fine tints, 61; flat stippled, 62, 104; graded tints, 104; line tints, 61, 104; ruled tints, 104; solid colors, 62, 104
Type: density on wax-engraved maps, 109; in European hand-lettering tradition, 114; foundry type, use for lettering, 64, 114; point size, 110; positioning on maps, 110; small sizes, 112; stamping by hand, 67; stamping by machine, 67; stamping stick, 65; technical peculiarities, 114. *See also* Lettering
Typographic etching, 44, 112
Typographic Etching Company, 43

Voisin, Russell, 78

Waite, Charles H., 28, 31, 121
Walker, Emery, 43
Warde, Beatrice, 106
Wax: building wax, 69; engraving wax, 51; fill-in wax, 69; ground-laying on case, 53; ground thickness, measurement, 54; halftonometer, 54; image transfer to, methods, 55–57; ingredients, 51; Levy halftone gauge, 54; molding compound, formulas and blending, 51–52. *See also* Case
Wax engraving: apprenticeship system, 39, 84; building, 67; building iron, 70; building pen, 71; building-up process, 71; burrs, 58; business orientation to, 90; capital equipment for, 86–87; cerography, 16–23;

checking, 72; color plates in, 120; in common printing press, 119; corrections in engraving, 67; in county atlases, 48; Dawson process, 43–44, 112; deep engravings, 54; double-line tools, 58; early contracts, 129–37; electrotint, 42; in England, 41–44; in England, economic unimportance, 47; in Europe vs. U.S., 44–48; in fire insurance maps, 49; flaming, 72; flaming tool, 72; fountain tool, 70; fusing, 72; in geography texts, 26–27; glyphography, 24, 41–44; graver, 58; hold, 69; medium resistance, 94; mender, 67; in newspapers, 116; origins, 11–19; picking up, 72; point symbols, 57, 95; in railroad maps, 31–33; relief line engraving, 31; replacement by offset lithography, 38; in road maps, 49; shavings, 58; singeing, 72; skilled practitioners as limiting factor, 39; specialist categories, 84; stylus for freehand lines, 98, 101; task, nature of, 57; thin engravings, 54; tones, 61–62; typographic etching, 43–44; in U.S., development, 23; experimental period (1840–50), 24; transitional period (1850–70), 27; main period (1870–1930), 28–38; replacement by lithography (1930–50), 38–41; in U.S. vs. Europe, 44–48; V-shaped tool for freehand lines, 98. *See also* Electrotyping; Engraver; Lettering; Lines; Lithography; Relief printing; Tones; Type; Wax
Willis, Nathaniel, 16
Woodcut, as relief printing process, 6–7
Wood engraving: cutting knife,

167